International Congress on Sustainability Science and Engineering (ICOSSE 2023)

Houston, Texas, USA
19-21 September 2023

ISBN: 978-1-7138-8679-2

International Congress on Sustainability Science and Engineering (ICOSSE 2023)

Houston, Texas, USA
19-21 September 2023

AIChE
The Global Home of Chemical Engineers

Printed from e-media with permission by:

Curran Associates, Inc.
57 Morehouse Lane
Red Hook, NY 12571

CURRAN ASSOCIATES INC. proceedings.com

Some format issues inherent in the e-media version may also appear in this print version.

AIChE
120 Wall Street, FL 23
New York, NY 10005-4020

Phone: (800) 242-4363
Fax: (203) 775-5177

www.aiche.org

Additional copies of this publication are available from:

Curran Associates, Inc.
57 Morehouse Lane
Red Hook, NY 12571 USA
Phone: 845-758-0400
Fax: 845-758-2633
Email: curran@proceedings.com
Web: www.proceedings.com

TABLE OF CONTENTS

Techno-Economic Analysis of an Integrated Approach for Bio-Methanol Production from Micro-Algae in the UAE

Authors

Al Lagtah, N. - Presenter, Newcastle University

In this proposed research, an integrated approach is taken to produce bioenergy from micro- algae, by modelling bio-methanol production through gasification of biomass using Aspen Plus. The key stages of the process are gasification, syngas post treatment, methanol synthesis followed by purification. The developed model performance is then evaluated for three different types of micro-algal species: (1) *S. Acuminatus* (2) *N. Oculata* and (3) *S. Almeriensi*. The performance of the micro-algal species was also compared with a plant biomass (pine). The simulation results were found to be comparable to experimental data available in literature.

S. *Acuminatus* was found to produce the greatest amount of bio-methanol and the greatest HHV value under identical conditions. A sensitivity analysis was conducted to evaluate the optimum gasification temperature for obtaining maximum bio-methanol yield, for each of the feedstocks. The optimum temperature was found to be 860; 880 and 860 °C for S. *Almeriensi*, N. *Oculata*, and S. *Acuminatus* respectively.

Since obtaining specific raw material costs for each micro-algal species was difficult, a general feedstock of *Chlorella vulgaris* is considered. A sensitivity analysis was conducted for wide range of bio-methanol selling prices and two *Chlorella vulgaris* feedstock prices ($269 and $45 per ton). The average bio-methanol selling price is $770 per ton. However, when applied as a shipping fuel, the cost of bio-methanol increases manifold to $3,336 per ton. Hence, the best scenario is when methanol is sold at the bunker prices. An IRR of 525% was recorded even when the raw material price was as high as $269 per ton. The worst case was identified to be when the bio-methanol price was $693 per ton. Minimum profitability was obtained with a payback time of 14 years (without 3-year construction) with a plant lifespan of 20 years. An IRR of only 2.6 % was obtained. Considering the standard average price of $770 per ton, an ROI of 2.6% and 13.2% was noted when raw material prices were $269 and $45 per ton, respectively. Therefore, the plant can be classified as feasible.

Sustainable Supersonic Aviation: Redefining Air Travel with Renewable Energy and Lift Fan Vtol Technology

Authors

Rabbi, M. F. - Presenter

Abdullah, M., University of the Kentucky

The aviation industry will need to drastically cut emissions to reach net zero by 2050, and this is projected to happen through carbon capture technologies and other measures. The aviation industry must keep working to enhance the performance of conventional aircraft to progress. Efficiency and sustainability can be greatly improved by introducing innovations in aircraft design, fuel efficiency, engines, and materials. The paper highlights the latest innovations in the aviation industry aimed at creating more sustainable and efficient aircraft designs. Specifically, the paper focuses on two key innovations: the use of renewable energy sources, such as solar and wind power, and lift fan VTOL technology. The former offers a solution to the problem of limited runway space by allowing supersonic passenger aircraft to take off and land vertically; the second alternative reduces carbon emissions, increases fuel economy, and lowers operational costs. The combination of these two innovations has the potential to revolutionise the aviation industry by making supersonic passenger aircraft more sustainable, efficient, and versatile. In this paper, we explore the technical details behind these innovations and their implications for the future of supersonic passenger aircraft.

Sustainable Biofuels for Aviation and Maritime Decarbonization

Authors

Tan, E. - Presenter, National Renewal Energy Lab

Reducing greenhouse gas emissions in the transportation sector is a major challenge of the 21st century. This sector is the largest source of emissions in the United States, responsible for one-third of all emissions. Meeting the goal of net-zero GHG emissions economy-wide in less than three decades (by 2050) requires a significant reduction in transportation GHG emissions. However, decarbonizing commercial aviation and maritime shipping is particularly difficult due to the inability to rely on electricity or hydrogen as alternative energy sources in the near or medium term.

Biofuels have the potential to play a crucial role in decarbonizing the aviation and shipping sectors and reducing their environmental impact. They are produced from renewable carbon sources and can directly replace traditional fossil fuels. Biofuels offer additional benefits such as reducing air pollution and improving energy security by diversifying fuel sources. However, there are challenges associated with using biofuels, such as finding sustainable feedstocks and potential competition with other sectors. Also, biofuel production depends on the feedstock type, which impacts sustainability. Therefore, it is vital to consider the environmental and social impacts of biofuel production and use and ensure sustainability criteria are met.

Biofuels derived from waste materials and non-food crops, such as catalytic fast pyrolysis (CFP) bio-oils, have the potential to reduce reliance on fossil fuels and create a circular economy. CFP bio-oils are renewable and have significantly lower life cycle greenhouse gas emissions compared to fossil fuels. They capture carbon from the atmosphere and store it in bio-oil and bio-char. Ultimately, opportunities for biofuels lie in their potential to provide a sustainable and low-carbon alternative to traditional fossil fuels for aviation and maritime decarbonization.

Food Security Challenges in Kashmir Valley India

Authors

Bhat, M. A. - Presenter, Model Government Degree College Chararisharief

Raja, V., Chandigarh University

Bhat, A. H., Model Government Degree College Chararisharief

Factors such as social, political and armed conflicts can become the cause of food insecurity and hunger in any area of the world, developed or underdeveloped. Their individual impact may be low and mitigatable, but their collective impact seems devastating. Kashmir valley in India is appearing as a hot-pot of food insecurity due to one of these factors. Political instability in Kashmir as a result of seven decades old armed conflict between India and Pakistan over Kashmir has made Kashmir a ripe place for food insecurity. Kashmir valley is surrounded by mighty Himalayas and has a very small land area fit for agricultural purposes, therefore its protection and development for agricultural purposes should have been the priority of the local Governments. This study depicts the otherwise. This study was conducted to evaluate the role of political instability in food insecurity in Kashmir valley. The results suggest that due to the corruption and favouritism at Government level, vast agricultural lands have been consumed for other purposes which could have been avoided. Developmental projects are not carried in a way that can guarantee the protection of agricultural lands. Haphazard and not well-thought engineering of these developmental works has destroyed more lands in the adjacent resulting in the destruction of thousands of hectors of prime agricultural lands between 2000-2023. The pace at which the trend of conversion of agricultural land in to other purposes is happening in Kashmir valley, it is estimated that by 2050 the Kashmir valley will guarantee food security for less than 50% of its population.

Industrial Scale Thin-Film Composite Membrane Modules for Salinity-Gradient Energy Harvesting through Pressure Retarded Osmosis

Authors

Low, M. - Presenter

The current work highlights development of novel thin-film composite hollow fiber membranes for pressure retarded osmosis applications to harvest salinity gradient energy. The membranes were developed with a specific target of harnessing the salinity gradient energy between wastewater reverse osmosis retentate and seawater reverse osmosis brine via osmotic mixing. The hollow fiber membranes were prepared by coating a polyethersulfone substrate with a thin-film composite polyamide layer via interfacial polymerization, which were assembled into modules of different diameters for lab scale and pilot scale evaluation. As the module size increased from 1-inch diameter. to 2-inch., and 4-inch., the water permeability, tested against a 1000 mg/L sodium chloride solution at 15 bar, decreased from 2.6 L m^{-2} h^{-1} bar^{-1} to 2.0 L m^{-2} h^{-1} bar^{-1}, and 1.2 L m^{-2} h^{-1} bar^{-1}, respectively. The power density, measured in the lab-scale unit using 1 M sodium chloride draw solution, and DI water feed solution, also decreased from 9.1 W/m^2 to 5.3 W/m^2 with increasing module size. Pilot scale evaluation of the 4- inch and 8-inch modules on a 24 m^3/day unit resulted in lower power densities of 2.5 W/m^2 and 1.5 W/ m^2 which translated to 0.024 kWh/m^3 and 0.05 kWh/m^3 of salinity-gradient energy harvested, respectively.

Reducing the Impact of Produce with Efficient Regional Controlled Environment Agriculture Facilities

Authors

Hilberg, G. - Presenter, Continuum Energy

Agriculture produces over 10% of the carbon in the US and its contribution continues grow. 30-40 percent of all food is wasted and a larger percentage of produce is wasted. Over 70% of the carbon impact of produce is from long distant transportation. Regionally located controlled Environment Agriculture (CEA) facilities (primarily greenhouses and vertical farms) have the potential to address both of these issues, but can have much larger on-site energy usage. Applying state of the art HVAC, lighting, controls and energy storage systems has the potential to allow regional CEA facilities to operate with low/no Scope 1 & 2 emissions and allow for short distant transportation which has a high potential for decarbonization. This presentation is based on a new build CEA facility (currently in construction) that is targeting operating as a zero carbon facility by 2025. The presentation will include details of design and engineering of the large scale heat pumps for heating & cooling, 100 % LED lighting, rainwater collection and other novel features that position this location as a carbon free solution for the produce industry.

Intermolecular Analysis of Ionic Liquid Mixtures Using Kamlet-Taft Parameters

Authors

Dahl, R. - Presenter

The American Innovation and Manufacturing Act of 2020 initiated the phase-out of hydrofluorocarbon (HFC) refrigerants for more environmentally friendly hydrofluoroolefin (HFO) refrigerants. In response, Project E.A.R.T.H (Environmentally Applied Research Towards Hydrofluorocarbons) and our collaborators are actively pursuing methods of recycling HFC refrigerants that will no longer be usable. However, widely used mixtures of HFC refrigerants often form azeotropes which are not separable through conventional separation techniques like distillation. Thus, new separation methods are necessary and ionic liquids are proposed as entrainers in extractive distillation for the separation of HFC refrigerant mixtures. To design these complicated new systems, thermophysical data for ionic liquids and ionic liquid mixtures are necessary but uncommon. Our group aims to elucidate the complex behavior of ionic liquid mixtures with respect to their thermophysical properties in hopes to examine potential correlations between thermophysical properties and Kamlet-Taft parameters. Kamlet-Taft parameters Ï€*, Î±, and Î² represent polarizability, hydrogen bond donating ability, and hydrogen bond accepting ability respectively. We hypothesize that the molecular characteristics described by Kamlet-Taft parameters are related to viscosity, density, and thermal conductivity. In which case, the necessary thermophysical data needed to design HFC separation systems could be derived from Kamlet-Taft parameters. This correlation would aid the effort to integrate ionic liquids as an entrainer in extractive distillation systems for the separation of hydrofluorocarbon refrigerant mixtures.

A CIRCULAR ECONOMY VISION FOR NEXT-GEN NUCLEAR POWER

Authors

Singh, V. - Presenter, Curio Legacy

The 94 nuclear reactors in the U.S. collectively produce about 20% of the nationâ€™s electricity at >90% capacity factors and with a stellar safety record. More importantly, nuclear accounts for 50% of Americaâ€™s carbon-free generation. However, compared to the lofty expectations of its pioneers, nuclear energy has unfortunately proven inadequate. Despite its reliability and proven scalability, the publicâ€™s perception of nuclear power remains less favorable compared to other zero-carbon power sources. Polls show that the issue of â€œnuclear wasteâ€� consistently ranks at or near the top of the arguments posed for objecting to the expanded use of nuclear power.

The politics surrounding the federal program to store nuclear waste in a geologic repository in Yucca Mountain, Nevada, has led to a spectacular impasse. As a result, the nationâ€™s inventory has grown to nearly 90,000 metric tons and is stored in some 75 locations across 35 states.

Unknown to many, however, is that â€œnuclear wasteâ€� or â€œspent nuclear fuelâ€� is neither *waste* nor *totally spent*. Technically speaking, it is *partially used nuclear fuel*. For instance, the energy content within the accumulated transuranics (or heavier-than-uranium elements) alone represents energy equivalent to 1% of the worldâ€™s natural gas reserves. Challenges related to economics and risks of nuclear proliferation have prevented commercial recycling of this material in the U.S.

With pollution, climate change, and the need to alleviate energy poverty all weighing heavily on the public conscience, nuclear power is in a resurgence with several proprietors of advanced reactor designs heading for commercial deployment. But their experience has shown that yesterdayâ€™s nuclear waste is holding back tomorrowâ€™s nuclear reactors from gaining public acceptance.

NuCycleâ„¢ is Curioâ€™s next-generation process that recycles nuclear waste efficiently to recover an array of commercial products. With NuCycle coupled with its next-gen reactor design, HOPEâ„¢, Curio is laying the foundation for a long-awaited circular economy that will bolster the case for wider adoption of nuclear power.

Intellectual Property Considerations for Sustainable Innovations

Authors

Townsend, P. - Presenter, Finnegan, Henderson, Farabow, Garrett & Dunner, LL

As climate change continues to affect the world, industries of all kind are developing new ways to increase their sustainability and decrease their carbon footprint, in part through carbon capture technologies. This green revolution has led to a variety of new proprietary sustainable technologies being brought to the market. With the increase in new sustainable technology, the necessity for intellectual property continues to rise. Although companies need to protect their IP rights, there is also a need to allow other companies to help the world become more sustainable. Thus, companies must balance the need to protect IP rights and the need to share their technologies.

This presentation will provide an overview of the basics of IP, how IP is at the forefront of sustainable technology, and important considerations for members of industry, academics, and anyone with an idea for a sustainable invention.

Employment of Green Methodologies for Valorized Processing of Dairy Waste Products into Bioactive Peptides: Recent Advances.

Authors

Kehinde, B. - Presenter

Igiehon, O., Louisiana State University

Sharma, P., Pacific Northwest National Laboratory

Over the years, the global dairy industry has been critiqued for the green house gases it emits into the atmosphere. Commercial dairy farms have been studied to emit large amounts of greenhouse gases such as nitrous oxide, methane and ammonia especially through waste products generated in their courses of operations. Products such as animal offal, blood, and stool, and other processed derivatives such as whole milk, whey, cream, skimmed milk and protein concentrates are proficient substrates for these emissions. Conversely, several studies have shown that these materials bear sufficient proteins that can be hydrolyzed into peptides of biological functionalities. In addition, sustainable and eco-friendly procedures such as ultrasonication, high pressure processing, and microwave processing amongst several others have been investigated and reported to be potentially effective in proteolyzing protein isolates obtained from dairy wastes. Furthermore, peptides derived from such protein lyses have been found to be beneficial against several metabolic syndrome disorders such as hypertension, cancer, and diabetes. Scientific studies have proven that green food processes are employable for the proteolysis of dairy-derived proteins into simpler hydrolysate fractions of lesser molecular weights. Subsequent purification with techniques such as Sodium dodecyl-sulfate polyacrylamide gel electrophoresis (SDS-PAGE), Reverse phase HPLC (RP-HPLC), ultrafiltration and Ion exchange chromatography, are further used to purify the peptides and technologies such as Edman degradation and Matrix-assisted laser desorption/ionization-time of flight (MALDI-TOF) are applicable for sequencing of amino acids. *In vitro* examinations through biochemical assays and *in vivo* experimentations with laboratory mice have reported that these purified peptides indeed are potentially functional for the management of bodily disorders. The enhanced application of environmentally friendly techniques for peptide isolation can be proficiently optimized into continuous and cost-effective remedies for concerns related to dairy wastes.

Metal Organic Materials (MOM) Spheres for Dephenolization in Nanofiltration of Olive Wastewater

Authors

Njimou, J. R. - Presenter

Maicaneanu, S. A., Indiana University of Pennsylvania

Godwin, J., Kogi State College of Education (Technical)

Elambo, N. G., National Advanced School of Public Works, P.O. Box 510, Yaounde, Cameroon

Talla, A., Research Center, National Advanced School of Public Works, P.O. Box 510,

A Metal-Organic Materials (MOM) water filter blended in aluminum oxide was fabricated from secondary resource orange-peel powder and aluminum salt. This nanocomposite (MOM) was characterized by Fourier Transform Infrared Spectroscopy (FTIR), X-ray diffraction (XRD), Scanning Electron Microscopy (SEM) and Energy Dispersive Spectroscopy (EDS). MOM was used to elaborate alginate aluminum orange-peel beads (Alg/MON). A fixed-bed reactor of these spheres was manufactured and used for dephenolization in nanofiltration fraction of olive wastewater and recovering compounds of the interests for the future perspectives of manufacturing new products in cosmetic, pharmaceutical and food industries. It have demonstrated in this study that during the intermediate stage of operation in a column fixed-bed in bi-components system at a constant flow rate, the adsorbed compounds of interest were replaced by the influent phenol molecules due to the weaker affinity of compounds of interest for Alg/MOM. It resulted in outlet stream concentration of compounds of interest i.e, hydroxytyrosol/polyphenols effluent higher than its feed. Phenol exhibited a sequence: single < binary < multiple components system while hydroxytyrosol/polyphenols showed the sequence: single > binary > multiple components system. Alg/MOM allowed the separation of olive mill wastewater (OMWW) before phenol hydraulic retention time (HRT) as a supplementary of water resource. The recovering hydroxytyrosol/polyphenols from its HRT is up to 90% in its solution at purity can be operated for the effluent above 90% in bed volume (BV). Alg/MOM column was saturated above 90% with the feed and needed regeneration to restore its adsorptive capacity using 50% EtOH. The data were modeled with the good correlation from Thomas, Yoon-Nelson, and Adams-Bohart models, reflecting the dephenolization in nanofiltration and the regeneration of Alg/MOM spheres at desorption cycles except in the intermediate zone with superior hydroxytyrosol/polyphenols concentration in feed. The Metal-Organic Materials (MOM) spheres are suitable in agrofood wastewaters processing.

Keywords: *Metal-Organic Materials (MOM); Dephenolization, Hydroxytyrosol/polyphenols; Recovery; Hydraulic retention time (HRT); Modelling; Agrofood wastewaters.*

Rethinking natural fermentation to simultaneously decarbonize food ingredients, consumer chemicals and transportation fuels

Authors

Tracy, B. - Presenter, Superbrewed Food

The manufacturing of food ingredients, consumer chemicals and transportation fuels account for the majority of CO_2 emissions in the US. Interestingly, current manufacturing practices of food commodities, such as sweeteners and protein concentrates, commonly takes advantage of economies of scale by integrating with the production of consumer chemicals and transportation fuels. Accordingly, Superbrewed is innovating upon this same nexus approach by harnessing natural fermentation to simultaneously produce food ingredients, consumer chemicals and transportation fuel from corn, resulting in a net-zero or even CO_2 sink life-cycle.

At our core, Superbrewed discovers and ferments natural (i.e., non-GMO) microorganisms that are themselves high-performance solutions for markets with strong customer demand, and which produce biochemicals that are drop-ins to fossil-based chemicals and fuels. Our immediate focus is the commercial production of Postbiotic Protein and butyric acid. Postbiotic Protein is a holistically better food protein ingredient compared to animal proteins, as it delivers enhanced consumer experience in taste, texture, nutrition and sustainability across a broad range of foods. Butyric acid directly replaces coal-based butyric acid, in a range of applications such as flavors, fragrances, nutritional supplements and consumer goods like automotive and aviation coatings.

This nexus approach of harnessing natural organisms to maximize product and revenue outputs of anaerobic fermentation while delivering net-zero solutions to multiple industries also delivers improved economics for a broad range of biochemicals, sustainable aviation fuel and renewable natural gas.

Design, Fabrication and Optimization of an Inclined Continuous Flow Solar Water Distillation System to Treat Contaminated Spring Water.

Authors

Talinda Kasule, H. - Presenter, Kyambogo University

Obbo, O., Kyambogo University

The rapid growth of Ugandaâ€™s population over the past two decades has placed significant strain on the current national water supply service. While some communities in Uganda use the main NWSC water supply, umbrella authorities, NGOs and other private sector organizations as their main water source, other residents depend mainly on groundwater springs to access clean water. However, due to factors like inadequate spring protection and poor solid waste management, some of these springs have become contaminated with physico-chemical and microbiological pollutants. This research project focused on utilising solar water distillation as an alternative method for treating contaminated spring water in Masajja Village, Makindye Ssabagabo County, Wakiso, Uganda. A continuous flow, inclined solar water distillation system was designed and modelled using SolidWorks software, and fabricated using heat-strengthened glass, hardwood timber and mild steel as the main materials. The major methods used for fabrication were gas metal arc welding, and metal, wood, and glass panel cutting. The performance of the solar distiller was then analyzed by measuring hourly and daily distillate productivity, determination of optimum input flow rate, determination of performance efficiency, and physiochemical characterization of both the raw feed water and the distillate. From the experiments performed, it was concluded that the solar distiller has production efficiency of 36.88% and productivity of 4.71 litres/m^2, which can be sufficiently scaled up to meet domestic water needs. The solar distiller temperature and inlet flow rate also have a direct influence on the productivity of the solar distiller, with high distillate volumes being observed at higher temperatures (75.5°C) and moderate flow rates (437ml/min).

Novel EV. Its Estimations for Comparison

Authors

Judbarovski, D. - Presenter

davidjud.blogspot.com/2023/05

davidjud.blogspot.com/2023/05

Novel EV. Its estimations for comparison

EV with breakthrough powering and charging

David Judbarovski, Israel, principle inventor, judbarovskidavid@gmail.com,

Viktor Tkachenko, prof., DSc., director of SPE RESST NSC KIPT, tkachenko.vikiv52@gmail.com

Abstract

The EV can be charged in a few minutes for 650 km mileage, and powered by device 80 kW, 100 kWh, being of 110 kg (250 liter totally, 160 liter at 300 bars), USD 3200.0 if being mass produced, anymore less than USD 80.0/kW.

Background

Typical battery of EV is about 80 kW, 100 kWh, for 650 km mileage, and 550 kg, 900 liter, USD 30.000 now, because lithium being needed for it is a bottleneck in supply for EV-industry, so its market price is crazy skyrocketing. It needs some hours for full charging, and lifetime is lower 5 years.

Short disclosure

A car powering on-board device consist of a fire resistant hermetic capsule containing both a fuel cell and a tank with pressed hydrogen, divided one from another by a fire resistant wall too. The capsule has two inlets: for pressed hydrogen and for atmospheric air, and two outlets: for water vapor exhaust and electricity for the car powering. It is quite hazard free hydrogen usage for mobile powering. Different hazard free design for hydrogen was published earlier for energy supply for housing

Distributed charging stations are designed, either by electrolysis, or by concentrated rays of magnetrons, for the water splitting in hydrogen, collected in a small buffer storage tank at 300 bars for the cars fueling. If 3 cars in 6 minutes fueled simultaneously, it is 8000 kW peak electricity consume, can be renewable electricity.

A Systems Perspective for Navigating the Foundations of Sustainability and a Circular Economy

Authors

Meidl, R. - Presenter, Baker Institute For Public Policy

Global energy production, resource demands, heightened environmental and social concerns, and the growing waste crisis have spurred an increased focus on sustainable strategies and a circular economy. With a global and national focus on a climate plan that prioritizes clean energy, it is critical that future actions and policies incorporate life cycle dimensions across value chains and understand how energy transition planning can impact the larger sustainability agenda. Absent from discussions are life cycle management of materials and wastes and social equity and environmental justice of energy transitions. These issues greatly influence the true sustainability profile of companies, governments and investments.

The Feasibility of Palm-Based Biomass Co-Firing and Different Oxygen Carriers to Enhance the Performance of Coal-Based IGCC Power Plant

Authors

Al Lagtah, N. - Presenter, Newcastle University

The constantly growing global energy demand for energy is still produced from fossil fuels. Even with the rise in the global awareness about the negative effects non-renewable resources of energy, the abundance and the low cost of coal still ensures the existence of coal-fired power plants.

This research investigates the potential of retrofitting coal-based IGCC power plants with palm-based biomass co-firing and chemical looping combustion (CLC) carbon capture for zero carbon emissions. The investigated biomass are palm coir, palm empty bunch and palm leaflets. The effect of different biomass co-firing and CLC oxygen carriers on the techno-economic performance of coal-based IGCC power plants was modelled and simulated using Aspen Plus software.

The results showed that the use of iron oxide as an oxygen carrier resulted in better performance than nickel oxide in terms of total plant efficiency, total plant energy output and carbon specific emissions. However, to achieve zero carbon emissions, the flow rate of the oxygen carriers was 15 times the flow rate of the feed.

Palm coir co-firing resulted in better performance regardless of the used oxygen carrier. However, the performance has diminished when palm empty bunch and palm leaflets co-firing were used. The presence of particulate matter results in downstream equipment fouling and hence, lowering the overall power plant efficiency. This issue has been addressed by adding cyclones for syngas cleaning.

The use of CLC for carbon capture has increased the cost of power generation depending on oxygen carrier type. The cost has increased by 32% and by 45% when iron oxide and nickel oxide were used, respectively. However, there was no clear trend to indicate if increasing biomass percentage will increase or decrease the cost of power generation. Unfortunately, the lowest cost was greater than actual market trading prices. Therefore, currently no clear economic drive for developing coal and biomass co-firing IGCC power plants with CLC carbon capture technology. Even though, a satisfactory co-firing IGCC power plant with suitable decarbonisation scenario might be economically feasible, depending on a trade-off between the decarbonisation degree and CO_2 specific emissions.

Study of the Structure Interfacing of 2-Dimensional Materials for Optoelectronic Devices

Authors

Nasr, N. - Presenter

Inorganic cesium lead halide ($CsPbX_3$) together with 2D transition metal dichalcogenides have indicated an improvement in optoelectronic performances of optoelectronic devices. We have broadly examined the structural, electronic, and optical properties of $CsPbI_3$ with two-dimensional MXY (M = Mo, W, and XY = S, Se), by using DFT calculations.

The bandgap of the isolated surfaces and the heterostructures between $CsPbI_3$ and two-dimensional MXY were calculated. The effective mass of isolated surfaces with the $CsPbI_3$ and 2-dimensional MXY heterostructures was calculated and the differential charge analysis were executed. Fermi level and Eigenvalues (VBM/CBM) were calculated for the heterostructures and are presented in the study.

The simulation results show that the bandgap was reduced when $CsPBI_3$ was coupled with other heterostructures. The analysis indicates that $CsPbI_3/MoS_2$ has the greatest absorption in the visible-ultraviolet range while the absorption of MXY surfaces is lower than $CsPbI_3$ bulk.

All results indicate that interfacing $CsPbI_3$ with MXY effectively improves absorption spectra in the visible-to-ultraviolet region, and thus $CsPbX_3$ when coupled with 2D transition metal dichalcogenides has a great potential to improve the efficiency of 2-dimensional materials in optoelectronic device applications.

Making Water Sexy: A Collaborative Approach to Innovation and Technology

Authors

VandenHeuval, K. - Presenter, The Water Tower

When it comes to advancing the water industry, protecting our communities, and recruiting the next generation of water workers, innovation and collaboration go hand in hand. With up to 50% of water workers retiring in the next 5-10 years, innovative technology is not only crucial to streamlining and improving treatment and processes for utilities, but also necessary to attract and engage the future water workforce.

The Water Tower, the new global water innovation hub northeast of Atlanta, Georgia, is breaking down traditional silos in the industry and bringing together utilities, universities, private industry, and NGOs under one roof to expedite implementation of innovative water and wastewater solutions. TWT offers resources including a living laboratory with live water process flows, state-of-the-art laboratories, and a network of water experts for the industry to conduct pilot testing and R&D. Hands-on training programs and technology demonstration days allow utilities to explore and operate innovative technologies firsthand.

Several innovative digital and physical technologies currently demonstrating their products at The Water Tower will be highlighted, including a smart septic system, smart hydrant, acoustic leak detection system, and more.

By forging powerful R&D collaborations, fostering economic development, and equipping water and wastewater utilities with the research and technology they need, The Water Tower strives to make a positive impact on the industry and our communities.

Advancing Research for Climate Risk Assessments, Adaptation, and Mitigation in the Water Sector

Authors

Samples, S. - Presenter, The Water Research Foundation

Climate change is altering our natural hydrologic cycle, creating uncertainty when it comes to the quality and quantity of water sources available to municipalities. There is a need for the water sector to take action to meet their communityâ€™s water needs. The Water Research Foundation (WRF) pursues climate change research to help equip utilities with the knowledge and resources they need to combat this uncertainty. WRFâ€™s research takes a holistic, and One Water approach to climate change. This presentation will provide a high-level overview of the three subtopics of climate change research WRF conducts. We will cover climate risk assessment, climate adaptation, and mitigation strategies. At the end of the presentation, the audience will have a better understanding of what types of climate change research is available at WRF, how they can access it, and what opportunities exist to participate in new climate change research.

An Integrated Process for the Production of a Sustainable Solid Biofuel As a Drop-in Replacement for Coal in Power Boilers

Authors

Satyavolu, J. - Presenter

Torrefied (mildly pyrolyzed) biomass represents a high quality renewable and carbon neutral energy source. Based on pilot and demonstration scale trials by universities, research labs, and companies, the torrefied biomass briquettes are shown to be the â€œdrop-inâ€™ replacement for coal in power plants. However, the inherent constituents such as volatiles and ash present in biomass and overall cost of the briquettes are challenges to the performance and economic viability of the torrefied briquettes. Integration of a xylose isolation process developed by University of Louisville prior to torrefaction is shown to improve the quality and performance of the torrefied briquettes as well as improve the economic viability of the torrefied biomass briquettes. In this presentation, we discuss the integrated process, improvement in quality and performance of the torrefied briquettes, as well as a coproduct opportunity. This work is done in collaboration with BioProducts, LLC, Louisville, KY.

AI Enables Industrial Autonomous Operations

Authors

van der Bijl, E. - Presenter, Yokogawa Corporation of America

According to industry surveys, a growing number of manufacturing organizations are adopting artificial intelligence (AI). AI has grown into a broad category that comprises several technologies such as computer vision, machine learning (ML), and natural language processing (NLP).

Commanding the hype most recently have been generative AI applications such as ChatGPT, which build upon NLP to analyze and generate language-based data such as text and speech. Although this technology appears poised to revolutionize industrial operations, it is far less applicable to industrial process control.

Meanwhile, another AI technology, reinforcement learning, has proven successful for direct control of a plant. In March of 2023, ENEOS Materials Corporation and Yokogawa announced an agreement in which Factorial Kernel Dynamic Policy Programming (FKDPP), a reinforcement learning-based AI algorithm, will be officially adopted for use at an ENEOS Materials chemical plant. The agreement followed a successful field test in which a solution, â€œautonomous control AIâ€� demonstrated a high level of performance while controlling a distillation column at the plant for nearly a year. This is the worldâ€™s first example of the adoption of reinforcement learning AI for direct control of a plant.

Experts in the manufacturing and process industries have estimated that more than 65% of process control loops are underperforming and up to 30% are operating in manual mode. For companies striving toward autonomous operations with little or no human presence at plants, underperforming controls and manual operations have emerged as major issues.

The AI solution controls distillation column operations that were beyond the capabilities of existing technologies such as proportional-integral-derivative (PID) loop control and advanced process control (APC). That had necessitated manual operations based on the judgements of experienced plant personnel.

This session will describe the autonomous control AI development process and how it addressed the challenges presented by difficult and complex process operations. The autonomous control AI eliminated off-spec production and reduced costs in terms of feedstocks, fuel, and labor. The technology also reduced steam consumption and CO_2 emissions each by 40%. Ultimately, it enables alignment between management and operations as well as industrial autonomy.

Scalable, Transportable Thermochemical Energy Storage for Industrial Process Heat and Steam on Demand

Authors

Dwivedi, A. - Presenter

Energy storage based on reversible chemical reactions, termed thermochemical energy storage (TCES), has been proposed since the 1970s. TCES based on calcium oxide and hydroxide, where dehydration of Ca(OH)2 to CaO stores energy and the hydration of CaO to Ca(OH)2 releases the energy, is especially promising. The material is cheap at $100/ton, has high energy density at ~500 (WÂ·h)/kg, the reaction is highly reversible, and the reaction temperatures of >550 C can serve ~75% U.S industrial process heat demand. Additionally, the temperatures are high enough to drive efficient heat-to-power conversion cycles for electricity generation. However, the commercialization of TCES has been hindered due to: 1) a large heat exchange surface requirement that accounts for ~80% of heat storage system cost, making it expensive and non-scalable and 2) storage material challenges of agglomeration, and low heat and mass transfer. Cacheâ€™s innovative system solves both problems by using high-strength CaO pellets flowing through a continuous fluidized-bed reactor (FBR). Since the storage material is fluidized and flows through, instead of being packed inside the reactor, the system can be scaled to 100s of hours of continuous energy storage/supply without scaling up the reactor. The FBR has ~10x higher heat and mass transfer compared to conventional reactors, reducing the heat exchanger surface required and hence the system cost significantly down to $ 5-10 /kWh. The CaO pellets have high strength and prevent fine particle generation and thus agglomeration. Once charged with waste heat or renewable electricity, the pellets store energy at ambient temperature and can be easily transported to other sites where energy is needed. The proposed highly scalable energy storage system can dispatch energy continuously for >100 hours while storing energy at ambient conditions, insulation free for up-to > 6 months with near zero loss. The technology enables energy transportability while providing high temperature process heat/high quality steam for ~75% U.S industrial demand

Structure-Property Relationships in Intercalated Graphene Oxide Membranes for Separation of Salts and Hydrocarbons from Petroleum-Containing Wastewaters

Authors

Sonker, M. Jr. - Presenter, RGIPT

Graphene oxide (GO) membranes are emerging as a viable new technology for separation of complex aqueous streams. Prior work by the Nair group and collaborators has developed scalable GO membranes that a robust in a variety of harsh operating environments. Additionally, they demonstrated that GO membranes intercalated by conjugated polyaromatic molecules showed enhanced rejection of salts as well as excellent rejection of organics (including hydrocarbons). However, the mechanisms underlying these findings are not well understood. The aims of my work are threefold: (1) Determine the key aspects of the microstructure-separation property relations of GO nanofiltration membranes that are intercalated/functionalized by a representative p-conjugated polyaromatic intercalant toluidine blue O (TBO). A combination of experimental techniques will be used for this purpose. (2) Based on initial insights from aim 1, synthesize an expanded set of â€œGO-Xâ€ membranes intercalated with a systematically chosen set of polyaromatic intercalants (â€œXâ€), and refine the structure-property relationships with a focus on enhanced rejection of monovalent salts such as NaCl. (3) Apply these fundamental advances to develop GO membranes for petroleum wastewater recycling. Large amounts of wastewater are generated in petroleum refineries from crude oil desalting operations (~80 million metric tons in the US). This wastewater contains significant amounts of crude oil, as well as salts such as NaCl (which is extracted from crude oil by washing with water). These wastewaters are commonly sent for biological wastewater treatment and are not reused. Detailed separation measurements on these wastewaters with the engineered GO-X membranes, and use of the process data in conjunction with ASPEN flow sheeting, will allow the development of a membrane-based process for production of reusable water and recovery of valuable crude oil from the desalter effluent. Preliminary work on all three aims shows encouraging results that set the stage for achieving the aims of this research.

Comparison of Household Organic Waste Management Systems in the European Union: Best Practices and Constraints

Authors

Celestino, E. - Presenter

Carvalho, A. I., Instituto Superior Técnico

Palma-Oliveira, J. M. Sr., University of Lisbon

Rapid and uncontrolled population growth, urbanization and modern lifestyle have put cities under pressure. The large production of urban waste is one of them. In the improvement process of household urban food waste and the progressive transition toward a circular economy, according to the goals established in Directive 2008/851, from January 1st, 2024, all European Union (EU) countries may opt for the separation and recycling of biowaste at source or through the selective collection. As a result, EU countries are under pressure to implement or restructure their household organic waste management systems to comply with this directive, and the practices they apply can be replicated in other places of the world. However, the rate of separate collection of household organic waste has been quite heterogeneous, with some EU countries reaching higher levels than others. Therefore, this article aims to contribute to the improvement of household organic waste management practices, aiming at the transition to a circular economy, based on the definition of the best practices applied by the member states, which can be replicated in other states. A Case Study methodology has been applied in this work, where 49 reports from organizations of all EU were analyzed, covering the 27 state members case studies. Based on this analysis, a roadmap will be proposed with information that will be valuable for policymakers and professionals in strategic areas to improve household organic waste management practices. In this roadmap, it will be possible to identify factors such as socioeconomic-demographic characteristics, the type of collection and treatment, restrictions on landfills and communication campaigns that can impact the success of implementing a closed-loop waste management system. This means a system, which seeks to ensure quality recovery, reduce the use of virgin resources, keep this waste out of landfills and move towards a circular economy. As a result of this work it was identified that in the most developed countries in organic waste management, household organic waste management systems are based on four pillars: legal restrictions on the use of landfills, regular communication campaigns, selective door-to-door and mandatory collection and well-defined fee methodologies.

Challenges in Achieving Carbon Neutrality & Plastics Circularity by 2050 – An Engineering Perspective

Authors

Dhodapkar, S. - Presenter

Industrial decarbonization and plastics circularity are major challenges of the 21st century to achieving a sustainable future. As both a recognized leader in the production of olefins/polyolefins and a major user of energy, Dow has a key responsibility. By 2030, Dow has committed to reduce its net annual carbon emissions by 5 million metric tons vs its 2020 baseline (15% reduction). By 2050, Dow intends to be carbon neutral (scopes 1-3 plus product benefits). Dow is conducting research to develop new process technologies as well as its integration with existing technologies to drive the production of olefins to net-zero emissions.

Similarly, the need to develop sustainable solutions for plastics circularity is an urgent challenge for the planet. Dow has committed to accelerating the circular ecosystem by transforming waste and alternative feedstock to deliver 3 million metric tons per year of circular and renewable solutions by 2030. Dow has proposed a two-pronged approach: â€œStop-The-Wasteâ€� and â€œClose-The-Loopâ€�. Additionally, significant product innovations are in the pipeline to design for recyclability.

As we are refining our strategies and scale our pipeline to deliver circular economy solutions by focusing on circular materials, technologies and value chain ecosystems, this presentation will highlight the challenges ahead. Scale, scope, rate of implementation, uncertainty, reliability and economics of decarbonization pathways will dictate our success. Decarbonization and circularity are inextricably linked, therefore a holistic view is imperative. Further, implementation, adoption and proliferation of transformative technologies will require supportive policy and financial incentives.

Mindful Manufacturing - Link between Economics and Sustainability

Authors

Ahmed, A. - Presenter

Many factors are pivoting chemical companies in the direction of sustainability and decarbonization. increasing environmental regulations, advancements in green chemistry, consumer demands for sustainable products along with long-term economic benefits are some of the catalysts that are propelling chemical companies to operate in an environmentally conscious manner.

Carbon capture, greenhouse gases emissions reduction, low-carbon feedstocks and energy conservation efforts are some of the initiatives that are getting embedded into the operating strategies of several chemical companies. Implementing such strategies can give quick first-hand results and also help strategically place manufacturing facilities ahead in the â€˜green race.â€™ However, to realize long-term substantial reduction and net-zero targets, radical changes in the operation of chemical companies are required. Such radical approaches warrant consideration of several factors including but not limited to 1) the Production of products with greener feedstocks, 2) Design changes in facilities and 3) Investments in R&D to create innovate sustainable products. As such, these efforts are money intensive and hence result in higher costs that are passed down to the consumers.

Such efforts raise many questions: Are chemical companies becoming overly ambitious with goal setting to achieve zero carbon footprint? Is true sustainability achievable and quantifiable? Are consumers willing to pay a premium for greener products? Can short-term losses with investment in such initiatives be recovered with long-term benefits? How are such efforts defining the operation of chemical facilities? What are the current and future economic outlooks of companies taking the green initiative? This talk aims to discuss the current efforts to achieve sustainability in chemical industries and attempt to find answers to the above questions.

Adsorptive Removal of Heavy Metals and Organic Dye Simultaneously from Complex Wastewater Using Recyclable Magnetically Chitosan Functionalized EDTA Adsorbent

Authors

Kim, H. - Presenter, University of Seoul

Verma, M. - Presenter

The existence of more than one class of pollutants in wastewater is a challenging task to treat via one-step process due to significant differences in physicochemical properties. Here, we prepared a new and effective adsorbent magnetite-doped chitosan-ethylenediaminetetraacetic acid (Fe_3O_4@CS-EDTA) for the simultaneous removal of multiple heavy metals: Hg^{2+}, Cd^{2+}, and Ni^{2+}, and methyl orange (MO) dye from complex wastewater through adsorption process. In this adsorbent, we believe that EDTA in the synthesized adsorbent work as a cross-linker as well as a chelating agent for the capturing of heavy metal ions, while the protonated amino groups of CS adsorb MO through electrostatic interaction. In the single component system, the isotherm data followed to Langmuir model and gave the maximum adsorption capacities (q_m) of 232.70±14.30, 121.40±7.0, 56.50±2.20, and 732.10±72.40 mg g^{-1} for Hg^{2+}, Cd^{2+}, Ni^{2+}, and MO, respectively. Kinetics data strongly fitted to the pseudo-second order (PSO) models, confirmed the chemisorption process. Interestingly, in the case of the binary component system containing both metals and MO, the adsorption capacity of the adsorbent for the dye was not affected by metal presence, while the adsorption of heavy metals was increased with increasing MO concentration. elemental mapping, XPS and FTIR confirmed the adsorption mechanism via complexation and electrostatic interactions. Moreover, no significant loss in the adsorption efficiency even after the six continuous adsorption-desorption cycles confirms the great stability and potential of the adsorbent in treating the complex wastewater.

Sustainable Energy and Materials Development

Authors

Krishnamoorti, R. - Presenter, University of Houston

The development of sustainable paradigms for energy and materials are crucial to net-zero targets. While renewable electricity and bio-renewable materials are critical enablers for this transition, there are numerous bottlenecks that require technological breakthroughs and trade-offs. Two technological developments might contribute substantially in this arena: development of scalable carbon dioxide removal technologies and advancement in polyolefin circularity. Through a discussion of these advancements, I will discuss an â€œall of the aboveâ€� strategy to achieve a sustainable energy and materials paradigm.

Remaining Useful Life Estimation of Proton Exchange Membrane Fuel Cells Using Artificial Neural Network

Authors

Biswas, M. - Presenter, University of Texas At Tyler

Biswas, M., Tuskegee University

Salim, M., University of Texas at Tyler

Wilberforce Awotwe, T., King's College London

With the recent investigation into the possible replacement of fossil fuel commodities due to their harmful effects on the environment, fuel cells today are being recommended by the research community as the suitable option, particularly in the automotive industry. Proton Exchange Membrane Fuel Cells (PEMFC) are energy converting devices that transform chemical energy into electrical energy with the byproducts of water and heat. With relatively high efficiencies and virtually no greenhouse gas emissions, fuel cells are often ideal for the automotive applications. Many studies and investigations have been carried out to explore ways to improve the overall performance of the fuel cell as well as to minimize the fuel consumption. Despite the progress made by the research community in developing various predictive models in order to address the performance issues, the accuracy of these developed models has lately become active research direction. Due to the complex and nonlinear behavior of the multiple fuel cell variables, the research has focused on modeling using artificial neural network (ANN), which can account for a system of multiple variables with nonlinear and complex behavior to fit a model and estimate the performance at low error and high convergence rate. The current study explored the accuracy of artificial neural network with varying neurons and learning algorithms in predicting the remaining useful life of a PEM fuel cell. Comparing the mean square error (RMSE) and coefficient of determination (R^2) of the models, the developed models yielded the relatively low error values (<0.25) and high R^2 values (>0.9) indicating a higher accuracy. The ANN based model shows promising results in the prediction of the PEMFCâ€˜s dynamic performance and at different operating variables. In addition, the ANN model can be used to estimate the for automotive applications to effectively power efficiency and remaining useful life.

Investigation of Plasma-Induced Nickel Nitride Formation during Ammonia Synthesis

Authors

Lin, Z. - Presenter

Caron, D., Princeton University

Zheng, Y., Princeton University

Kondratowicz, C., Princeton University

Koel, B., Princeton University

Plasma-assisted catalysis for ammonia synthesis provides a potentially more sustainable alternative to traditional thermal catalysis. Advancing the use of plasma with catalysts in tandem requires a more thorough understanding of plasma-surface interactions and reaction kinetics. In the present work, we investigate the formation of nickel nitride on a nickel film deposited on a stainless steel substrate in a low pressure N_2 dielectric barrier plasma discharge (capacitively coupled; 12 kV, 20 kHz). Surface morphology of the nickel film after 40 minutes of plasma treatment was examined by scanning electron microscopy (SEM). Energy-dispersive x-ray spectroscopy (EDS) and high-resolution x-ray photoelectron spectroscopy (HR-XPS) determined the formation of nickel nitride on the plasma-treated nickel surface. Depth profiling using XPS shows that the formation of a nitride extends beyond a surface atomic layer. To further understand the formation of nickel nitride in the presence of a N_2 plasma, a microdischarge cell capable of generating a N_2 plasma at 5-10 Torr inside an environmental scanning electron microscope (ESEM) while providing optical access for direct SEM imaging and EDS analysis was constructed. This ESEM cell enables surface analysis after plasma treatment without any distortion of elemental and structural information by air contamination. These initial results and additional studies will be helpful in understanding the role of nickel nitride in plasma-assisted catalysis of ammonia synthesis using nickel catalysts.

Artificial Neural Network Based Model for Reverse Osmosis Water Desalination Membrane

Authors

Biswas, M. - Presenter, University of Texas At Tyler

Salim, M., University of Texas at Tyler

Barakat, N., University of Texas at Tyler

Mossad, M., University of Texas at Tyler

Availability of fresh drinking water continues to be a major concern around the globe. One of the most effective methods to provide fresh drinking water is through the use of desalination plants. Over the past years, reverse osmosis (RO) membrane technology has developed significantly, gained popularity, and become the most widely used desalination technology for both seawater and brackish water, due to advancements in membrane technology and improvements in energy efficiency resulting in higher performance and reduced cost. On the other hand, the modeling of RO membranes is challenging due to the complexity of the analytic models and the lack of the membraneâ€™s characteristics. This work provides a multi-input-output Artificial Neural Network (ANN) model to predict the membraneâ€™s characteristics at different flow conditions. ANN is a biologically inspired computer program that is constructed to mimic the way in which the human brain handles information, and it has been used widely to model different systems especially nonlinear and/ or complex systems. This model can be used to reduce the modeling complexity of the membrane and to predict the freshwater flow rate and salinity at different conditions by using a set of experimental data for the membrane. The model uses the feedwater pressure, flowrate, and salinity to predict the freshwater flowrate and salinity. The model shows correction value (R)of 0.985 for the testing data and it is capable to predict the membraneâ€™s characteristics at pressure values from 90 to 160 psi.

Hydrogen Horizons in 2050 Energy Market: Advancements, Challenges, and Future Perspectives

Authors

Danish, M. S. S. - Presenter

This study offers an expansive exploration into the burgeoning domain of hydrogen energy, casting light on the current landscape, technological advancements, encountered obstacles, and future prospects. Amid a global shift towards cleaner, more sustainable power sources, hydrogen energy emerges as a promising contender, poised to revolutionize power systems. Yet, the path is not without hurdles. This study delves into the persistent challenges obstructing large-scale hydrogen deployment, from infrastructural constraints to technical barriers and economic feasibility. Moreover, this research proposes a novel framework to facilitate hydrogen energy's seamless integration into power systems, paving the way for a hydrogen-powered future. Drawing on exhaustive review and analysis, it charts a roadmap that navigates the complex terrain of hydrogen energy. This investigation offers a unique lens into hydrogen's potential and the mechanisms required to unlock it. It serves as a comprehensive guide for researchers, policymakers, and industry stakeholders navigating the hydrogen revolution.

Impact of Substrate Makeup on PFAS Elimination Rates from Sludge, IX Resin and AFFF Using Supercritical Water Oxidation

Authors

Viswanathan, S. - Presenter

Deshusses, M. A., University of California, Riverside

Nagar, J.

Rising concern of Perfluorinated alkylated substances (PFAS) contamination of our ecosystem has sparked interest in this pollutant as it pertains to water and waste management. While PFAS sources are numerous, it is widely believed that firefighting foam and extensive industrial uses are common pathways for PFAS compounds to proliferate into our ecosystem.

In an attempt to better understand the fate and transport of PFAS compounds undergoing supercritical water oxidation (SCWO), a one (1) wet ton per day scale SCWO system was employed to study the elimination efficiencies of this process. There is a pressing need to develop and validate advanced treatment technologies that can destroy PFAS in a variety of substrates.

In this paper, we report on how three distinct PFAS waste substrates from three different sources were treated using SCWO process. The waste includes lime stabilized sludge from a municipal wastewater resource recovery facility; aqueous film forming foam (AFFF) from a DoD facility; and spent ion exchange (IX) resin from a 'pump and treat' water treatment facility.

Supercritical water oxidation or SCWO for short, is a physical-thermal process that relies on the unique reactivity and transport properties of water above its critical point of 374 °C and 218 atm. At these conditions, organics are fully soluble in supercritical water, and with the addition of oxygen, all organics rapidly and completely oxidized to form carbon dioxide, clean water, and inorganic salts. The studies examined a variety of PFAS compounds, both targeted and non-targeted, but most specifically focused on PFOA and PFOS. SCWO, on average, was able to eliminate 99.95% of PFOA and 99.99% of PFOS across all waste substrates, and greater than 99.9% elimination of all other PFAS compounds combined. Non-targeted PFAS was accounted for using laboratory scale verification tests by employing fluorine mass balance. No hydrogen fluoride was found in the effluent gas, and all the fluorine from the destroyed PFAS was accounted for as fluoride in the effluent water, with no low molecular weight or volatile PFAS compounds in the emission.

A Hybrid Methacrylate-Norbornene-Thiol Photopolymerization System for One-Pot Synthesis of Adhesives with Tunable Physical and Mechanical Properties

Authors

Alrefai, M. - Presenter

Maric, M., McGill University

The photopolymerization of a hybrid methacrylate-norbornene-thiol system is employed to synthesize adhesive formulations with adjustable physical and mechanical properties. The suggested one-pot synthesis offers an environmentally friendly, zero waste approach for the rapid synthesis of UV-curable solventless formulations. The combination of methacrylate and thiol-ene brings together the advantages of both free radical homopolymerization and thiol-ene step growth reactions. Thiol-ene reactions offer benefits such as delayed gel point, high conversion, low shrinkage and stress, and tolerance to oxygen. This study investigates the impact of various factors, including the thiol:ene ratio, thiol functionality, and the addition of a methacrylate as a hard monomer, on the physical properties as well as the viscoelastic and adhesive performance of these formulations. Initially, pressure-sensitive adhesives (PSAs) are produced when ethylene glycol dicyclopentenyl ether methacrylate (EGDEMA) is crosslinked with the di-functional thiol (EBM) to create crosslinked networks with high conversion (up to 100%) and low glass transition temperature (Tg=-2.5°C to -13.5°C). Various thiol:ene ratios are studied; rheological studies indicate that these formulations make PSAs exhibiting best adhesive/cohesive performance at thiol:ene ratios of 0.5-0.75. Elastomeric polymer films are then produced when EGDEMA is crosslinked with a tetra-functional thiol (PEMP) resulting in highly crosslinked films with high conversion (up to 100%), higher Tg (ranging from 9°C to 17°C), and increased gel content (up to 100%). Finally, temporary adhesives are produced by the incorporation of a hard monomer, isobornyl methacrylate (IBOMA) along with a bifunctional thiol (EBM), resulting in thermoplastic polymers with low gel content and variable Tg (-12°C to 41°C).These findings suggest the potential utilization of this hybrid system as one-step UV-curable formulations to produce polymer networks for various applications, depending on the materials used. The environmental benefits of this approach include eliminating the production of waste and by-products, as well as eliminating the need for heat, solvents, or any post polymerization processing such as purification and precipitation.

Mechanism-Guided Design of Lignin Material for Sustainability

Authors

Yuan, J. - Presenter

Li, J., Washington University in St. Louis

Li, Q., Texas A&M University

Hu, C., Texas A&M University

Ragauskas, A., University of Tennessee

Lignin is the second most abundant biopolymers on earth and can serve as a low-cost and abundant feedstock for sustainable biomaterial. However, lignin remains primarily a waste stream from paper, pulping, and biorefining industries, due to the lack of fundamental understanding of structure-property relationship between lignin chemistry and material properties. Throughout the past decade, our interdisciplinary team has substantially advanced such understanding by defining how lignin molecular weight, uniformity, linkage profile, and functional groups could impact the properties of the lignin-derived sustainable materials, including carbon fibers, nanoparticles, recyclable plastic blends, and pavement materials. In particular, we have found that higher molecular weight, better uniformity, and more linear linkages could improve carbon material properties, empowering quality lignin carbon fiber for energy and environmental sustainability. Based on these fundamental understandings, we have recently advanced a new chemical design to substantially increase lignin molecular weight, ether linkage content, and hydrophobicity. The design was achieved through anhydride modification to produce a new type of lignin, high-molecular weight esterified linkage lignin (HiMWELL). The HiMWELL design achieved more uniform esterified linkages, larger molecular weight, and much lower â€"OH group *via* inter-unit linkage. These chemical characteristics perfectly overcome the inherent chemical limits of lignin, including the diverse chemical linkages, heterogenous functional groups, variable molecular weights, and amphiphilic structures. As a result, HiMWELL lignin has exhibited superior properties as a precursor for various functional materials. First, HiMWELL-based carbon fiber achieved record tensile strength and >25% enhancement of mechanical properties as compared to polyacrylonitrile (PAN) carbon fiber. Mechanism study revealed the substantially increased crystal content in HiMWELL-based carbon fibers, as compared to those of carbon fibers from PAN and untreated lignin. The results highlighted that HiMWELL has the potential to serve as an additive to enhance biomaterial performance. Furthermore, we explored HiMWELL lignin as a guest polymer for plastics blends to improve recyclability. Unlike most lignin to decrease plastic blend performance, HiMWELL uniquely improved the mechanical properties of the recyclable polymer blends and achieved UV-shading effects. Overall, HiMWELL design overcame the inherent chemical structure limits of lignin and turned lignin into an enhancer for biomaterial performance for broad applications.

A New Carbon Economy: Innovating to a Post Pollution Future

Authors

Haggstrom, J. - Presenter, Lanza Tech

We currently recycle metals, plastics and paper - so why not recycle carbon?

There is an abundance of carbon locked in wastes from agriculture, forest, unsorted, unrecyclable municipal wastes and in gaseous byproducts of certain manufacturing processes. Biotechnology enables us to convert carbon pollution into everyday products, turning our carbon problem into an economic opportunity keeping the skies and oceans clean and blue for all!

LanzaTech has developed a gas fermentation technology that recycles carbon rich industrial off gases from many industries into fuels and chemicals. LanzaTechâ€™s carbon transformation solution leverages synthetic biology and AI to expand the product portfolio beyond the building block ethanol to other commodity chemicals that currently come from petrochemical feedstocks and underpin the material economy.

The potential for ethanol to be used as blend stock in gasoline has been well established for over a century. Through partnerships with consumer-facing companies such as Unilever, Lâ€™OrÃ©al, lululemon, and the Mibelle Group, LanzaTech has demonstrated the opportunity to use ethanol as a platform CarbonSmart™ intermediate in the production of goods including fragrances, cleaning products, plastics for packaging, and fibers for clothing. By recycling carbon into ethanol rather than starting with virgin fossil fuels, the production of everyday products can benefit from a lower carbon footprint.

This presentation will address the scaling up of carbon recycling technology and what it takes to cross the Valley of Death as an innovator!

Evaluating the Environmental Footprint of Jatropha Biodiesel Production and Utilization in Ethiopia: A Comprehensive Well-to-Wheel Life Cycle Analysis

Authors

Worku, Y. A. - Presenter

Ethiopia is exploring biofuels as an alternative to petroleum-based vehicles to address energy security and sustainability challenges. The biofuel program focuses on producing biodiesel from jatropha plants, which can reduce fossil energy consumption and greenhouse gas (GHG) emissions per kilometer driven when blended with diesel.

This study aims to assess jatropha biodiesel production using life cycle assessment and a well-to-wheel analysis, measuring energy balance and environmental impacts for 1 MJ of fuel. Inventory analysis is employed for field survey data, and scenario analysis considers planting conditions, regions, allocation methods, processing, and usage. The study also analyzes the energy savings, emission reduction benefits, and criteria air pollutants when blending different biodiesel-diesel ratios.

The results show that jatropha biofuel can reduce environmental impacts by up to 45-87%, depending on the type of irrigation used. The GHG emissions of 1 MJ of jatropha-based biodiesel produced in Ethiopia are 19.8g CO2 eq/MJ. When examining the global warming potential (GWP) of jatropha planting, it was found that the cultivation of jatropha accounted for the highest share of greenhouse gas emissions at 57.48%, followed by the biodiesel production process at 23.72%. On the other handB20 vehicles, with 20% biodiesel and 80% diesel, could save 17% of fossil energy use and reduce 13% of GHG emissions per km, compared to pure diesel vehicles.

The study's results unequivocally demonstrate that jatropha biodiesel has the potential to mitigate Ethiopia's transportation environmental impact. However, improving jatropha cultivation, oil extraction, and biodiesel production is necessary to achieve this potential.

Scale-Up of PureCycle Solvent Dissolution Recycling Process

Authors

Weber, M. - Presenter, PureCycle Technologies

PureCycle uses a patented, ground-breaking recycling process to separate virtually all color, odor, and contaminants from polypropylene waste feedstock to transform it into Ultra-Pure Recycled (UPR) resin.

This presentation will outline the evolution of the PureCycle process from the initial concept to commercial reality. A benchtop pressure cell was used to develop the fundamental polymer solubility data that supports the overall design concept. The main unit operations for the process were developed using a lab-scale batch system. Successful operation of the pilot scale system demonstrated that the purification process can be executed continuously. The commercial scale implementation uses a modular design that will enable rapid expansion of the process to multiple sites. The key challenges and successes at each stage of development will be discussed.

At every scale, the purification process has been demonstrated to produce recycled plastic with a color and clarity equivalent to virgin material. Analysis of the purified pellets has also confirmed the ability of the process to remove several substances of interest typically introduced to recycled plastic during the use phase. Purified plastic has also been used in blending studies to demonstrate its equivalence to virgin plastic in several product applications.

This process helps close the loop on the reuse of recycled plastics while making recycled polypropylene more accessible at scale to companies desiring to use a sustainable recycled resin. PureCycleâ€™s UPR resin is a like-virgin material that is easily colorable, and its production is projected to consume 79% less energy and produce 35% less carbon emissions than new polypropylene manufacturing.

Overview of Solid-Liquid Separation and Process Drying Technologies for Lithium, Battery Materials and Recycling

Authors

Perlmutter, B. - Presenter, BHS-Filtration Inc.

As the lithium and battery materials production and recycling increase in scalability, there is a need to understand solid-liquid separation and drying technologies for automated operation, improved efficiencies and reliabilities and safety. At each step, requirements change, and plants must ask critical questions for selecting the optimum technologies for producing high quality materials. In addition, there is a rapid proliferation of technological advances and cross- disciplinary nature of the work worldwide. The operating companies need â€œKey-Shapedâ€ engineers who have several areas of expertise with varying degrees of depth. Finally, there are increasing time and technology pressures to produce battery grade material from various sources as well as from recycling of production scrap and lithium-ion batteries. This presentation will provide guidance to engineers who are facing critical decisions for selecting process filtration and drying technologies for the various complex steps in battery materials manufacturing and recycling. The information presented will allow engineers to develop creative process solutions to achieve battery grade quality and a reliable and cost-effective operation.

A Low-Energy Process for the Production of Pure Dry Xylose from Sustainable Agricultural Biomass

Authors

Satyavolu, J. - Presenter

Next to glucose, Xylose is the most abundantly available sugar in nature. It is a low-calorie natural sugar that does not metabolize in the human body and does not show any increase in the blood sugar level. Commercially, for food applications, it is mostly available as a modified sugar alcohol, xylitol. Xylose is also a sustainable carbon source for biochemicals and biofuels - but, mostly these conversion reactions are carried out in aqueous phase reactions. Conn Center at University of Louisville (U*of*L) developed a low-cost and low-energy process for the production of pure dry xylose from sustainable lignocellulosic biomass. This process produces dry xylose under ambient process conditions using C5-rich hydrolyzates from biomass. Under a license from U*of*L, BioProducts, LLC is producing xylose from various cereal straws, grain hulls, and other biomass sources. The xylose isolation process using biomass hydrolyzates is discussed during this presentation.

Green Hydrogen to Gas (GH2G) Concept According to UN Agenda for Sustainable Development Goals-a Realistic Computational Modelling Based Approach for Chemical Industry

Authors

Zaidi, D. S. Z. J. - Presenter, University of Punjab

Asif, H., University of Punjab

Khan, R. U., University of Punjab

The development of sustainable approach is much needed for the 21st century technological solution in heavy chemical industries like petroleum refinery and fertilizer industries for urea production. NH3 is an important chemical specie which served as a raw material for urea production. Around the world NH3 synthesis took place via grey route e.g via methane processing which not only adds carbon footprints in the environment but also a great source of global warming phenomenon. In previous year catastrophic flood hit across Pakistan rural and urban areas which displaced millions of people and cause a loss of billions of $ to economy. This study focused on the computational modelling method for green production of Ammonia gas and is valuable for the stakeholders in order to take descions regarding decreased carbon footprints and adding sustainable growth over the network of heavy Chemical industries in Pakistan. Two important plants including AEC and SOEC are designed which showed efficiency of convertor for allied gases of upto 85% with additional production of excessive energy of 1155GW/Day.

Circularity & Climate Neutrality: Decarbonizing Products, Processes and Supply Chains

Authors

Skorpenske, R. - Presenter, Covestro

Climate change, population growth, unsustainable lifestyles â€" surmounting global challenges like these require embracing sustainability and the circular economy as a guiding light. A globally leading high-tech polymer manufacturer, Covestro has set a vision to be fully circular. Although the chemical industry has been an enabler for our modern, safe and sustainable life today, it also has to be transformed. In order to realize this transformation, the development of new technologies, materials and processes has gained increasing importance within the chemical industry. Covestro is exploring numerous opportunities including design for circularity, advanced recycling, mass balanced products and renewable energy solutions all of which will also help achieve their corporate goal to become climate neutral and achieve net zero emissions by 2035 for scope 1 and scope 2. In end, how is all of this possible? A key to help achieve climate neutrality and the circular economy is through strong partnerships and collaborations up and down the value chain as well as an innovative corporate culture.

Navigating the Selectivity Toward a Single C2 Product in CO2 Electroreduction

Authors

Gupta, S. - Presenter, Vale Technologies

Hoying, M.

The rational design of gas diffusion electrodes (GDEs) and catalysts play a vital role in achieving the industrially relevant performance metrics (e.g., selectivity, productivity, stability, and energy efficiency) of CO_2 electroreduction for a specific C_{2+} product. Along the reaction pathway for C_{2+} products formation over Cu-based catalysts, a crucial step of C-C coupling through dimerization of adsorbed *CO or coupling of *CO with its hydrogenated derivatives (*e.g.,* *CHO) has been identified by both experimental observation and theoretical calculation. This phenomenon provides an opportunity to design cascade reactions through CO intermediate to improve the yield of C_{2+} products. Some bimetallic tandem catalysts integrating Cu with a CO-generation metal (e.g., Ag and Au) have demonstrated the tandem catalysis effect, improving C-C coupling kinetics by increasing the partial pressure of CO (P_{CO}) near the Cu surface. However, when the tandem catalyst is applied in the conventional GDE comprising one homogeneous catalyst layer (CL), the CO utilization efficiency for C-C coupling is low because the local CO concentration throughout the CL is at the lowest level. Distinguished from the tandem catalysts, we introduce tandem electrodes, where the temporal and spatial CO concentration profiles are *in-situ* managed, to enhance the utilization of CO and thus drive cascade CO_2àCOàC_{2+} conversion with a yield at an industrial scale. We demonstrated that segmented tandem electrodes, even incorporating the commercial Cu nanoparticles, could reach over 90% Faradaic efficiency (FE) of C_{2+} products (60% FE for C_2H_4) at a partial current density of above 1 A cm^{-2}. We modulated the Cu surface by single-site doping to stabilize desired selectivity-determining intermediate, from which the formation of C_2H_4 versus C_2H_5OH branches. As a result, the selectivity toward C_2H_4 versus C_2H_5OH was controlled relatively. When the surface-modified Cu catalyst was integrated into the segmented tandem electrodes, the FE of C_2H_4 could be further increased to 70% at a partial current density of >1.5 A cm^{-2}. This presentation demonstrates the collective control of rate-determining and selectivity-determining steps to direct CO_2 conversion to a specific C_2 product at high production rates.

Net Zero and Process Safety

Authors

Sutton, I. - Presenter

Many companies in the energy and process industries have committed to reducing their greenhouse gas emissions. In many cases they have adopted the target of â€˜Net Zero by 2050â€™. The companies plan to reach this target by through a combination of emissions reductions and carbon capture and sequestration programs.

The effort required to meet this goal is, to say the least, formidable. If industrial managers are to meet the extraordinarily ambitious goals that they have set for themselves then they will need input from a wide range of professions. For example, the financial community has developed and and is implementing greenhouse gas emissions programs in many countries. In the United States the Securities and Exchange Commission (SEC) has published a proposed Climate-Disclosure rule that, if implemented, will require companies to report on (1) their current greenhouse gas emissions, (2) their programs for reducing those emissions, and (3) an assessment of the financial risk to which they are exposed due to climate change.

This paper will discuss how the principles of process safety management can be applied in order to help achieve this goal. The process safety community, although small, has an opportunity â€" even a responsibility â€" to make a contribution to climate programs. Areas in which its expertise can be helpful include (1) understanding the nature of high consequence/deferred risks, (2) making systems resilient, rather than efficient, (3) identifying and risk-ranking hazards, (4) understanding which elements of the traditional process safety management programs from OSHA and other agencies best fit a Net Zero program, and (5) maybe most important of all, the critical importance of systems thinking..

Coffee Point Wind

Authors

Noden, P. - Presenter

Coffee Point Wind (CPW) is a conceptual 8-12 + megawatt wind farm with solar subsidy stretching from Dillingham to Manokotak along the Nushagak Coast. This particular stretch of land is suitable for capturing the coastal wind off the Nushagak Bay and would involve three Village Corporations. CPW benefits include electricity sales, renewable energy careers, education incentives, and access to gorgeous land for tribal members with an access route to a landing/boat launch near the mouth of snake river. CPW would also provide a grid tie between Dillingham and Manokotak.

CPW will require buy-in and a strong partnership between the power producer Nushagak Cooperative, and three tribal corporations- Choggiung Limited, Saguyak Incorporated, and Manokotak Natives Limited. I am currently reaching out to all three Tribal Corporations who are eligible for federal and state grants and technical assistance through the Bureau of Indian Affairs, Department of Energy and the Renewable Energy Fund. Nushagak Cooperative will be able to buy power from CPW and sell it to residential customers utilizing Power Cost Equalization and a Power Purchase Agreement will need to be negotiated.

The project site is entirely on Native land which increases its value as revenues will go to village corporations and ultimately back to the community through dividends. Developing Coffee Point Wind will be a strong move toward tribal sovereignty because it provides our tribes the ability to produce and deliver power to our shareholders and descendants. Coffee Point Wind can ultimately allow us to heat our homes and businesses with electricity generated from renewable energy and offer us a form of energy security.

Investigation of Energy Harvest from the Thermal Decomposition of Munitions Wastewater

Authors

Adhikari, R. - Presenter

Parziale, N., Stevens Institute of Technology

Parziale, N., Stevens Institute of Technology

Su, T. L., Stevens Institute of Technology, Center for Environmental Systems

Christodoulatos, C., Stevens Institute of Technology

Braida, W., Stevens Institute of Technology

Fabrication processes in munitions plants produce an Ammonium Nitrate (AN) laden wastewater stream that needs to be disposed of in accordance with regulatory guidelines. This stream is composed of oxygen-rich nitrates, and carbon- and hydrogen- containing compounds, and therefore offers an opportunity to repurpose it as a nitrogen based monofuel for energy harvest. The thermal decomposition of small batches of the wastewater and control samples of comparable compositions at pressures from 0.1 to 10 MPa are explored in this work. The energy and emissions released from the decomposition of the samples are measured using a Differential Scanning Calorimeter (DSC), and a Fourier-Transform Infrared (FTIR) Spectrometer respectively. Experimental results are compared with the theoretically determined energy contents and ideal decomposition products to evaluate the â€œqualityâ€ of decomposition, and profiles and quantities of heat flow and emissions are used to elucidate major modes of sample decomposition. This investigation demonstrated that higher pressures improve energy release and reduce the amounts of harmful NO_x emissions. The decomposition enthalpy of the wastewater improved from 0.2 MJ/kg at 0.1 MPa to 1.8 MJ/kg at 10 MPa. Similarly, while NO and NO_2 emissions accounted for 6.5% and 2.5% of sample nitrogen, respectively, at 0.1 MPa, they accounted for only 0.7% and 0.08% the sample nitrogen at 10 MPa.

Application of Ultrasonic Separation in Bioethanol Purification Process

Authors

Liu, J. - Presenter, University of Illinois Urbana Champaign

Pearlstein, A., University of Illinois

Feng, H., University of Illinois at Urbana-Champaign

Bioethanol is a key renewable energy source that is now used as a gasoline additive and viewed as a promising feedstock in synthesizing renewable gasoline/diesel or sustainable aviation fuel. Annual U.S. production exceeds 16 billion gallons, mainly produced from the fermentation of starch or lignocellulose, with ethanol (EtOH) mass fractions in the broth ranging from 1-15% depending on feedstock, enzymes, and operating conditions. Water, as the major broth impurity, must be removed, and the significant energy consumption in purification becomes a major economic hurdle. State-of-the-art purification technology requires 30-60 % of the 29 MJ/kg EtOH combustion heat, depending on the EtOH broth concentration. Here, we describe the development of a novel fully electrified liquid/vapor ultrasound separation based on surface enrichment. The electrically-driven ultrasound transducer generates cavitation bubbles in the liquid, which rise in an ultrasonically-generated fountain, and shatter into fine droplets upon reaching the interface between the bulk liquid and headspace. Compared to distillation, no energy is consumed in effecting a liquid-to-vapor phase change, and second-law losses associated with heat transfer are greatly reduced. This approach also avoids the azeotropic bottleneck of atmospheric pressure distillation and the separate steps necessary to overcome it. Ultrasonic separation is influenced by bulk liquid temperature, carrier gas flow rate and inlet pressure, and transducer voltage. With the partial recycling of large drops (which are less enriched in EtOH than small droplets), the ultrasonic separation was significantly improved. The strategy of single and multi-stage separation investigated the feasibility of replacing the beer column and rectifier used in conventional distillation. Energy analysis indicates that ultrasound separation in bioethanol purification can significantly reduce energy consumption in dewatering. The potential annual energy reduction is about 0.5 quadrillion BTU for the U.S. and 0.7 quadrillion BTU globally. A carbon-balance analysis shows a potential annual carbon reduction of about 200 million tons.

WATER-ENERGY NEXUS:DYNAMICS, INTENSIFICATION, AND REUSE

Authors

Rosso, D. - Presenter, University of California, Irvine

Education and experience have formed us to foster steady-state conditions as the operational mode of choice for numerous water and wastewater treatment processes, including water reuse. However, due to the influent variability and the heavy reliance on electricity during processing, the energy in water reuse is almost never a steady-state operation. Furthermore, due to the dynamics of power generation, the carbon footprint of both power generation and water processes are almost never at steady-state. We present here our work on the water-energy nexus, process intensification, and resource recovery. We also discuss carbon and energy footprint analyses from case studies around the world.

Exxonmobil Lower-Emission Fuel Technology

Authors

Freer, A. - Presenter

Energy demand for transportation is projected to rise by 30 percent through 2050, and biofuels, hydrogen-based fuels and carbon capture and storage offer lower-emission solutions for hard-to-decarbonize sectors according to ExxonMobilâ€™s 2022 Energy Outlook. ExxonMobil Technology and Engineering Company (â€œExxonMobilâ€) has been researching and developing a range of renewable, lower-emission alternatives to traditional transportation fuels. Here we will present an overview of a selection of the products and process technologies we are evaluating and discuss the challenges and opportunities. For example, our ExxonMobil Renewable Diesel (EMRD™) process addresses the challenges for deep dewaxing with our BIDW™ catalyst which has been demonstrated to have the highest selectivity of multi-branched isomers. Another example of technology that we are developing is our Methanol-to-Jet (MTJ) technology for potentially converting renewable methanol into Sustainable Aviation Fuel. We also partner with companies, universities, and national labs to enable longer-term technology development spanning from feedstock through process and product. A recent example is our partnership with Toyota which demonstrated the compatibility of lower emission research fuels in existing passenger vehicles. Through these partnerships, ExxonMobil is focused on the fundamental science and engineering of advanced biofuels that are affordable, sustainable, and scalable.

Creating Lower-Carbon Technology Solutions for Society's Needs

Authors

Jones, E. N. - Presenter, ExxonMobil Research and Engineering Company

Energy is an essential component of modern living. Access to safe, reliable, and affordable energy is a critical enabler of higher living standards that include longer and healthier lives. Future energy demand will depend on global population growth, a rising middle class in developing countries, and how efficiently consumers use energy. All energy sources are needed, but meeting this demand is coupled with the need to reduce the risk to the environment, including climate change. In independent net-zero carbon scenarios, society needs to meet its energy demand with fewer, or even negative, emissions.

Power generation and road transport are the most familiar demands for energy, but over half of energy use occurs in the industrial sectors where the worldâ€™s goods and infrastructure are produced. Industries like steel and cement, along with commercial aviation and marine transportation, are difficult to decarbonize and will require technology innovations that can be deployed across a large scale. The level of innovation required to meet societyâ€™s net-zero ambitions is unprecedented. Partnerships between industries, governments, and universities will be essential to find multiple pathways to reduce energy-related emissions. ExxonMobil is actively developing biofuels, hydrogen-based fuels, and carbon capture and sequestration to offer lower-carbon solutions for the hard-to-decarbonize sectors. This presentation will share some insights into the challenges and opportunities for these new technologies, which need to be economically viable and executable at scale.

Globally Resolved Life Cycle and Techno-Economic Assessment of Algal-Based Biofuels

Conference

International Congress on Sustainability Science Engineering ICOSSE

Year

2023

Proceeding

2023 The International Congress on Sustainability Science & Engineering (ICOSSE)

Group

Abstract Submissions

Session

ICOSSE '23 Track B - Energy

Authors

Quiroz, D. - Presenter, Colorado State University

Greene, J. M., Colorado State University

Limb, B., Colorado State University

Quinn, J., Colorado State University

Advanced algal biofuels have the potential to support the decarbonization of hard-to-electrify transportation fleets. However, techno-economic analyses (TEAs) and life cycle assessments (LCAs) of algal biofuels have often focused on locations in suboptimal latitudes for algal cultivation, which can underrepresent the sustainability potential of the technology. This study performs a global TEA and LCA quantifying the regional economic potential and environmental impacts of algae biomass production and conversion to fuels across 6,685 global locations. A dynamic process model serves as the foundation for regionally resolved LCA and TEA to concurrently evaluate the techno-economic viability and environmental impacts of the system. The LCA accounts for the life cycle impacts associated with global electricity, hydrogen, and nutrient supply chains across 10 different environmental impact categories. Similarly, the TEA considers location-specific capital expenses, labor costs, and tax rates. The model predicts optimal annual average areal productivity values between 24.8 â€" 27.5 g m^{-2} day^{-1} in equatorial regions. In terms of LCA outputs, the global warming potential (GWP) of the system was found to be the lowest in South America and Central Africa with net emissions ranging between 55 - 70 g $CO_{2\text{-eq}}$ MJ^{-1}. Although GWP results show to be lower than the petroleum baseline (87 â€" 97 g $CO_{2\text{-eq}}$ MJ^{-1}Â), direct land use change emissions restrict algal biomass cultivation in these high-productivity regions. Additional environmental impacts such as respiratory effects and eutrophication potential indicate that algal biofuel production might lead to more detrimental effects than conventional diesel with optimal results ranging between 5 â€" 25 mg $PM_{2.5\text{-eq}}$ MJ^{-1} and 3 â€" 62 mg N_{eq} MJ^{-1}, respectively. Results from the TEA follow productivity trends and favor locations in Central America and Southeast Asia with minimum fuel selling prices ranging between $1.8 â€" $2.0 per liter of gasoline-equivalent. The presentation will present global maps of results that illustrate visually optimal places for algal production in terms of productivity, economics, and environmental impacts. Discussion will focus on understanding the challenges and opportunities to improve the economic viability and minimize the environmental impacts of algal-based biofuels.

Bio-Coal Production from Torrefaction of Coconut Husk with Food Waste As Binder: Effect of Pellet Size and Binder Ratio

Authors

Yadav, S. - Presenter, SNU

Nair, M. - Presenter

Coconut husk, a by-product of the coconut industry, holds promise as a sustainable feedstock for bio-coal production. However, pelletizing coconut husk poses challenges due to its low moisture content, low bulk density, and high ash content. To overcome these challenges, this study explored the use of food waste as a binder and investigated the effect of pellet size and binder ratio on bio-coal production from torrefaction of coconut husk. The study conducted experiments with three different pellet sizes (16 mm, 20 mm, 28 mm) and varying binder ratios (25%, 50%, 75%). Torrefaction was performed at 290°C for 60 minutes. The results demonstrated a clear relationship between pellet size and the properties of the resulting bio-coal. Smaller pellet sizes exhibit higher energy density and heating value, with the 16 mm pellet achieving the highest higher heating value (HHV) of 20.26 MJ/kg and an energy density of 1.25. Conversely, increasing the pellet size from 16 mm to 28 mm leads to a decrease in energy density and heating value. These results were corroborated by a Van Krevelen plot which revealed 16 mm pellet to have the lowest O/C and H/C ratio and was in vicinity with moderate quality natural coal.

The quality of the bio-coal produced was assessed through various testing and analysis techniques, including thermogravimetric analysis (TGA), proximate analysis, lignocellulose analysis, Fourier transform infrared (FTIR) spectroscopy, and field-emission scanning electron microscopy (FESEM). FESEM analysis clearly showed the presence of large pores in coal from 16 mm pellets, a very desirable property for any coal and FTIR showed the absence of OH, CO, and COOH groups that led to hydrophobicity of the bio-coal, another desirable property. In conclusion, use of two different biomasses in different roles, one as main feedstock, and other one as binder, makes this work very novel and provides alternative to effective waste management. The findings in this work provide insights for developing efficient and sustainable waste management processes, contributing to the utilization of vast waste biomass for renewable energy generation.

Green Hydrogen Worldwide Impact to Unlock a Carbon-Free Future

Authors

El-Maghraby, M. II - Presenter

Green hydrogen is environmentally friendly due to its zero emissions which neutralizes the environmental impact and hence control climate change. This paper aims to throughout light on green hydrogen production method which is water electrolysis, green hydrogen applications, production opportunities worldwide and the challenges in pursuit of a green hydrogen economy.

Hydrogen can contribute to energy security by providing another energy carrier with different supply chains, producers and markets. Hydrogen is attracting more interest because it has energy-dense and low-carbon fuel. Green hydrogen can be produced using water electrolysis method where water decomposed into oxygen and hydrogen by passing a direct current through solar or wind energy which drives electrochemical reactions. Furthermore, green hydrogen production can be achieved using water thermolysis method where thermal energy is the driving energy. Thermal energy thermolysis water steam is brought to temperatures of over 2500 K at which water molecule decomposes thermally.

The expected reduction in greenhouse gases is about 60% to 80% by 2050 through using hydrogen as pioneered energy efficient alternative power source to vehicle engines. Low variable renewable energy electricity cost is considered one of the opportunities for green hydrogen production with the continuously decreasing costs of solar PV from 250 $/MWh to 56 $/MWh and of wind electricity from 75 $/MWh to 48 $/MWh. There are a few barriers to be knocked down before the commercial aspect of green hydrogen is advanced; the cost of renewable energies needs to be viable itself as green hydrogen produced using electricity from renewable energy plant would be two to three times more expensive than grey hydrogen. Lack of dedicated infrastructure where only about 5 000 kilometres (km) of hydrogen transmission pipelines around the world compared with more than 3 million km for natural gas. Green hydrogen incurs significant energy losses at each stage of the value chain. About 30-35% of the energy used to produce hydrogen through electrolysis is lost. Storage of hydrogen as a liquid requires cryogenic temperatures which is very challenging.

Nutrient Optimization for Climate Smart Intelligent Food Supply Chain Dynamics Modeling

Authors

Donovan, R. - Presenter

In this paper we will describe methodologies through which the information flows coupling human and natural systems can be characterized through emerging information theoretic concepts and that can be subsequently used describe the dynamics of the human food with a specific aim to optimize food supply chains for nutrient density. The overall goal of the effort is to describe a system through which we can establish models, mechanisms and mathematics that leverage the relationship between the thermodynamics of information and the thermodynamics of Maxwell to describe how complex systems are bound together through mechanisms related to exchanges of information and associated entropic gradients. These concepts will serve as the foundation of a proposed methodology for design, and deployment of data enabled decision support tools through which suppliers (i.e. farmers and food processors) and distribution networks (e.g., cold food chain) can effectively mitigate impacts associated with the extremes of climate change that are certain to accelerate the food security crises affecting many underserved communities. The key features of these methods are: 1) Intelligent agent enabled, adaptively generated Knowledge Graphs (KGs); 2) Application of KGs to establish nutrient density parameters across a broad spectrum of food systems (crops, soils, management practice); 3) Application of KGs to model the dynamics of and design sustainable supply chains; and 4) Use of these methods as a basis for creation of decision support tools to optimize supply chain dynamics for nutrient density in the context of climate change impacts. These impacts include short term disaster scenarios such as intensifying storms and wildfires as well as longer term impacts such as the problems of urban food deserts. The proposed system is specifically designed to align with California Climate Adaptation Strategy, Californiaâ€™s Fifth Climate Change Assessment, and the California Climate Solutions for the Natural and Working Lands Climate Smart Strategy, via development of advanced AI approaches to identification of effective agricultural practice (e.g., regenerative farming) as it relates food nutrition.

Hierarchical Phonon Scattering from Nano to Macro Scale in Ag-Nano/TiO2-Micro Particle Decorated p-Type Bismuth Telluride Bulk Composites Using Ultrasonication Method

Authors

Rawat, P. - Presenter

Kumar, A., S.V.N.I.T

Jae, H. Y., Konkuk university

Jong, S. R., Kyung Hee University

We study the thermoelectric properties of a p-type BST composite with Ag nano-particle decorated TiO_2 microparticles (US-Ag/TiO_2). The dispersion of US-Ag/TiO_2 particles, synthesized by an ultrasonication (US) method, into the matrix effectively decreases lattice and bipolar thermal conductivity, attributed to the scattering centers formed at nano and micro scales. The electron backscattering diffraction (EBSD) measurements have revealed smaller grain sizes within the BST composite when paired with the US-Ag/TiO_2 particle dispersion. These reduced grain sizes, alongside nano-particle decorated microparticles dispersed throughout the matrix, scatter phonons effectively from long to short wavelength phonons and subsequently decrease lattice thermal conductivity. While the power factors of the composites are reduced, significant suppression of lattice and bipolar thermal conductivity has led to an increase in the maximum zT value (1.22 at 350 K) for a 0.9 wt% US-Ag/TiO_2 particle dispersion within the BST matrix. This particle dispersion in the BST composite consistently demonstrates a high zT value across an extensive temperature spectrum, leading to an exceptionally high average zT_{avg} value (1.2 up to 450 K), which is superior to the other values from reported BST composites. Thus, this research indicates that the dispersion of nano-particle decorated microparticles within a thermoelectric material matrix can significantly improve thermoelectric performance, which has promising implications for practical applications in thermoelectric cooling and sustainable, economical energy harvesting technologies.

Transforming Waste for a Circular, Safe, and Low Carbon Future

Authors

Martin, J. - Presenter, Dow Chemical Company

The future of waste is rapidly evolving as value chains seek to identify opportunities to effectively collect, recycle and reuse materials in multiple market sectors. Thought leaders envision that waste, including both plastics and bio-based, will be viewed as a valuable resource that can be used to create new jobs and an effective set of ecosystems by which it is transformed into new materials or building blocks. Established processes such as mechanical recycling will continue to evolve with the implementation of technologies that effectively sort, clean, and convert back into feedstocks that approach the performance delivered by virgin plastic resins used today. Additional advances in recycling systems that convert feedstocks back into hydrocarbon or oligomeric materials will be a vital part of the ecosystems where the demand for food-contact approved materials or routes to materials employed outside of the packaging sector require more innovative developments. Across the material recovery and repurposing systems, low-carbon solutions will continue to be demanded as the concepts of circularity and low-carbon become more entwined in both policy and infrastructure investment. This talk will discuss the options available and considerations for global and regional implementation.

A DFT Computational Modelling for Organic Aromatic Based Compounds As an Electrode for Na+ Ion Batteries

Authors

Awan, D. J. - Presenter

Zaidi, D. S. Z. J.

With the growing demand for energy storage systems, relying solely on lithium-ion batteries is not a sustainable solution because of limited disposal and recycle systems. It is necessary to explore alternative battery technologies that are cost-effective and can be produced with ease. To fulfill the energy storage demands of the future, our focus is on investigating the potential of 3-BROMOCATECHOL and pyridine as electrode materials for sodium-ion batteries. 3-BROMOCATECHOL and pyridine possess properties that make them suitable anode materials for bio-batteries, lithium-ion batteries, and sodium-ion batteries, such as low cost, ease of synthesis, stability, and ability to undergo charge-discharge cycles. This study examined the suitability of 3-BROMOCATECHOL and pyridine as electrode materials for sodium-ion batteries by evaluating their bond lengths, adsorption energies, HOMO, LUMO, and HLG energies. The results indicate that both materials exhibit favorable adsorption energies and stable Na-ion adsorption. Specifically, the adsorption energy of sodium in 3-BROMOCATECHOL ranges from 21.62 eV to 30.07 eV, while that of pyridine is 16.53 eV. The calculated bond lengths suggest stable Na-ion adsorption on both materials. The electronic properties, including HOMO, LUMO, and HLG energies, were also found to be suitable for efficient electron transfer during battery operation. Overall, these findings provide valuable insights into the electrochemical properties of 3-BROMOCATECHOL and pyridine and their potential as electrode materials for sodium-ion batteries.

Separating the Loop in Chemical Looping Reformation: How a Two-Stage Reformer and Decoupling Alleviates Process Challenges.

Authors

Krumm, R. - Presenter, University of Utah

Chemical looping reformation (CLR) is a process that converts a carbonaceous feedstock (refuse, biomass, coal, or natural gas) into syngas using a metal-oxide oxygen carrier. The traditional concept of CLR uses moving or fluidized beds to fully convert carbon to syngas in a pair of reactors for the oxidation and reduction. Sensible heat from the oxidation of the metal-oxide provides heat to sustain the reforming reactions. Steam can be added to the reformer as a source of additional oxygen. While CLR promises to provide a pathway to sustainable syngas, complexities with the process and physical limitations of the metal-oxide oxygen carrier have hindered its commercialization.

A decoupled approach to CLR may provide a solution. Rather than relying on the sensible heat from oxidizing the metal carrier, electric heaters provide the energy to break down biomass in the pyrolysis zone of the reactor. Steam added to the reactor aids in the conversion of long carbon chains to shorter, volatile species around 750°C. Not all the carbon is converted and the addition of an oxygen carrier, in this case Fe_2O_3, supports additional conversion. Unreacted feedstock, now mostly unconverted carbon and ash, discharges from the reactor mixed with Fe_3O_4 and FeO. Iron species are separated from the solids product and are sent to a separate, uncoupled oxidation reactor, hence open-loop. Gaseous products continue to the 1200°C gasification zone further reacting with steam to produce syngas. Syngas further undergoes water-gas shift and cleanup to produce commodity chemicals.

A simulation study using Aspen plus examined the effect of changing the operational parameters of the process focusing on minimizing the electrical load per unit mass of hydrogen produced. Refuse derived fuel (RDF) is the assumed feedstock. Parameters varied include steam and Fe_2O_3 feed rates as well as temperature in the pyrolysis zone of the reactor. Other benefits of decoupling the process are examined such as:

1. The catalytic effect of iron during pyrolysis
2. Improved heat transfer due to iron particle disbursement
3. Reduction of sulfur species in the syngas (e.g. H_2S) with the formation of iron-sulfides
4. Simplified material handling

Assessing the Viability of Carbon Capture and Storage in Nigeria: Opportunities, Challenges, and Implications

Authors

Alabi, I. - Presenter

Carbon capture and storage (CCS) is a technology that shows promise in mitigating greenhouse gas emissions and addressing climate change. Despite some recent advancements in carbon capture and storage (CCS) technology in Africa, its implementation in Nigeria, a developing country with a high carbon footprint and increasing energy demand, remains limited. This can be attributed to several challenges, including insufficient policies, funding, and technical expertise. This project aims to evaluate the potential of CCS in Nigeria by examining the opportunities, challenges, and implications of its implementation. It will conduct an analysis of the geological, technological, regulatory, and socio-economic factors that influence the feasibility of CCS in Nigeria. Furthermore, the project will explore the potential benefits and risks of CCS, including its effects on the energy mix, the environment, and the economy. The project's findings will offer valuable insights into the feasibility of CCS in Nigeria and inform policy decisions on the country's energy and climate strategies.

The project will employ a mixed-methods approach, integrating both qualitative and quantitative research methods. The qualitative methods will encompass a literature review, expert interviews, and case studies of CCS projects in Africa and other part of the world. The quantitative methods will entail data sampling and analysis to evaluate the geological storage capacity and potential for $CO2$ emissions reduction using CCS in Nigeria. Additionally, the project will leverage stakeholder analysis to identify the interests, perceptions, and attitudes of various actors towards CCS.

The project aims to offer a thorough evaluation of the potential of CCS in Nigeria, along with its effects on the energy, environment, and economy. By analyzing the technical, economic, social, and policy obstacles to the implementation of CCS in Nigeria, the project will suggest practical solutions to surmount them. The project will also suggest recommendations for creating a CCS roadmap for Nigeria that takes into account the nation's distinctive context and challenges. The project's contribution to global knowledge on CCS will inform the international efforts to reduce carbon emissions and achieve the objectives of the Paris Agreement.

The Dual Challenge of Net Zero and Circular Materials

Authors

Powell, J. - Presenter, Shell International Exploration & Production

To address sustainability goals for climate and the environment, the world seeks to achieve Net Zero (fossil) greenhouse gas emissions by 2050, and a circular economy for plastics and materials. These goals require a transformative redesign of systems and infrastructure for providing energy and chemical products. Currently, more than 80% of energy is derived from fossil fuels without carbon capture and storage, and less than 10% of plastics are recycled. Despite their environmental impacts, an increase in supply of energy and chemical products is needed to enable all global populations to achieve an equitable standard of living. The trilemma of reducing the footprints of energy and chemicals despite increasing demand is one of the grand challenges of our time. Given the extent of resources and investment required, systems analyses complemented with life cycle assessment are critical for optimization and decision making. This seminar will examine the challenges and opportunities in clean energy generation, transport and storage, carbon management, and circularity in delivering societal sustainability goals through 2100.

PET Extraction from Waste Plastic Material to be Used As Industrial Coating.

Authors

Alwahibi, S. - Presenter, National University of Science and Technology

Alsalmi, S. - Presenter, National University of Science and Technology

Polyethylene terephthalate or as shortened to be PET is a component present in the plastic material, with using a tertiary recycling method using other chemicals to extract this component then spread into metals, the purpose is to create a coating material from waste plastic that is environmentally friendly and corrosion resistance a valuable two for two products. As chemical/petrochemical engineers the petroleum industries major facilities is made out of metals or metallic components, these component goes through a huge amount of flowrates, pressure, and temperature that it affects it original construction in the long term, which makes it useless from an economical point of view, therefore most companies are trying to dedicate their work to find alternatives or at least chemicals, components, or materials that is useful to protect the facilities for a longer lifetime, when using a polymer as a nonmetallic material it has a better properties and characteristics making it the best choice used as protective coating, major example is plastic, therefore extracting PET or as known as polyethylene terephthalate to be used a coating is the practical importance of our study. It was found that 2 gm of plastic with 99 ml if solvent for a duration of 50 minutes can extract polyethylene terephthalate, then immediately placed on the surface it forms a coating sticking to the requested surface, making it a successful experiment in extracting the components and using the searched methodologies for the purpose of extracting a metal coating.

Guiding Innovation By Tracking the Sustainability Performance of Microbial Products throughout the Product Development Phase.

Authors

Sukumara, S. - Presenter, Technical University of Denmark

Meramo, S.

Humanity has exceeded natureâ€™s carrying capacity which could result into irreversible changes to the environment and global Earth systems, partly due to the over exploitation of non-renewable resources. Therefore, industries worldwide are venturing into possibilities of transforming processes from unsustainable sources to sustainable feedstocks to produce chemicals, food, and other functional materials. This transition towards sustainable development requires microbial products with better sustainability performance. Measuring the sustainability performance of these products demands the deployment of a comprehensive assessment framework to quantify the impacts of the research at various product development phases.

This contribution provides perspectives on successfully integrating environmental sustainability and economic viability assessments to guide product innovation in microbial fermentation-based products. Efforts on quantifying the economic and environmental performance of bio-chemicals are analyzed early in the R&D to highlight impacts, challenges, and opportunities for guiding bio-based products. The research will showcase how a Life Cycle Assessment based impact assessment framework could be scaled-up to quantify the planetary impacts using the Absolute Environmental Sustainability Assessment framework. Being relevant to the bio-based materials community, methodologies and applied cases will be tested to quantify the sustainability performance of fermentation-derived products.

Novel Process for Robust and Sustainable Anaerobic Wastewater Treatment

Authors

Gangal, M. - Presenter

With resources becoming increasingly scarce, industries have significantly optimized their processes to improve overall plant sustainability. This has helped reduce the impact on the environment while also directly improving companiesâ€™ bottom lines, due to significantly reduced costs associated with energy and freshwater consumption and reduction of hauled waste surcharges.

Anaerobic technology has become a go-to method to optimize wastewater treatment within a compact footprint and reduced costs. The wastewater obtained from industrial processes such as brewing are particularly conducive to anaerobic treatment due to their high rbCOD content, which can further be captured as a renewable resource in the form of biogas. Traditional treatment systems use either aerobic technologies which require a large energy input to degrade the organic pollutants or conventional anaerobic technologies that require a high level of operational and maintenance input and are prone to performance failures due to influent variability.

The use of a novel anaerobic technology utilizing high surface area plastic biomass carriers helps address the issues by providing a stable matrix for keeping the anaerobic bacteria within the reactors helping achieve high loading rates, with high uptime despite flow and load variations, thereby reducing impact on plant operations and improving overall efficiency. Depending on the incoming wastewater strength and temperature, the high methane content biogas produced often has the capacity to power the entire treatment process at the plant, while the lower biosolids production associated with the anaerobic process helps reduce the disposal costs at the brewery. Further pairing the anaerobic process with ultrafiltration and reverse osmosis membranes based treatment helps polish the effluent to very high standards. This paper discusses examples of implementation of the above concepts at breweries where the effluent or a portion of it, is re-used within the plant.

The implementation of biomass carrier based anaerobic treatment technology paired with membrane filtration can thus help industries reduce their ecological footprint and greatly improve their sustainability metrics, by not only recovering resources from waste such as energy and water, but at the same time save on operational and disposal costs thereby having a direct positive impact on their bottom line.

Improved Approach for Benefit Analysis of Carbon Dioxide Reduction Methods That Use Rechargeable Batteries

Authors

Porter, L. - Presenter, Fluor Corporation

Deshotels, R. - Presenter

Abstract

Many methods and types of facilities have been proposed to reduce emissions of greenhouse gases. The choice of technology is usually determined by government organizations, rather than corporations. Scientific comparison of alternatives would ideally be done by cost-benefit analysis, taking into account who would pay the costs and who would receive the benefits. But there are problems with cost-benefit analysis in this situation. Government projects have a history of overrun budgets; making cost difficult to estimate. It is impossible to agree on the economic value of benefits (disaster avoidance) for events such as the disappearance of endangered species.

Benefit ranking is more useful comparison in this situation. Because there are not enough resources nor enough time to implement all carbon reduction technologies, ranking is necessary to accomplish the most benefit over time, especially because the benefits are cumulative. The ranked benefits can be evaluated using the sciences of chemistry, physics, thermodynamics, and statistical reliability, adjusted for the real-world effects of natural phenomena and human behavior.

The benefit of electrical storage batteries is compared for two design applications. The first is storage batteries close-coupled with solar power generation facilities, to extend the facility production to cover the peak electrical demand in the evening. The second application is batteries to power vehicles that will be recharged from the electrical grid. Battery effectiveness is measured as avoided carbon emissions per kilowatt of battery capacity. These particular applications of batteries were chosen because there is ample evidence of a world-wide shortage of the materials needed to produce efficient rechargeable batteries, and. because there are published data on hourly solar potential and peak demand.

The results showed that for the two locations studied, the pairing of batteries with solar power facilities is orders-of-magnitude more effective at avoiding carbon-dioxide emissions than using the equivalent battery capacity in electric vehicles. This is true regardless of the electrical energy sources available, and is insensitive to uncertainties in future battery technology.

Composite Linear Program Controls Ethane Load in a Typical Ethylene Plant.

Authors

Zhang, B. - Presenter

Maximizing the processing capacity of the plant is currently a common demand in ethylene cracking production. The accumulation of ethane in the system often causes the ethylene rectification tower to hit the load bottleneck. Timely and uniformly increasing the conversion rate of the ethane cracking furnace will fully make use of the capacity of the cracking furnace, reduce the ethane load of the ethylene rectification tower, and ensure that the impact on the run length of the ethane cracking furnace is minimized. Composite linear programming, as an option of the Aspen ethylene APC solution, can effectively feedback the ethane load information of the separation section to the advanced controllers of the ethane cracking furnace in the cracking section, and unify the adjustment strategies of these advanced controllers. Minimize the adjustment amount of each controller, reduce the impact, and relieve the ethane load of the ethylene rectification tower from exceeding the processing capacity.

The Energy-Carbon Nexus of Sustainable Energy Systems

Authors

Mutschler, R. - Presenter

To achieve carbon neutrality by 2050, Switzerland requires a deep understanding of its energy system and the potential for using carbon not only as an emission, but also as a recyclable commodity in Energy-X paths. In this work, we investigate the Energy-Carbon nexus using a MILP method based on Pyomo, implemented in Empas ehub modeling platform. We demonstrate the critical role of Energy-X in supplying sustainable fuels and balancing seasonal mismatches through storage. Integration of Energy-X results in seasonally optimized carbon capture and utilization and cost-effective storage solutions during times of limited import capacities or high prices. Our work provides valuable insights into achieving a net-zero CO2 emission energy system and is applicable to other national energy systems.

Fabrication of Novel Aerated Concrete Blocks By Combining Traditional and Synthetic Materials

Authors

Biswas, M. - Presenter, IIT Delhi

Bhaskarwar, A. N., Indian Institute of Technology Delhi

The demand for green construction materials is growing worldwide for environmental protection reasons. The majority of CO_2 emissions come from the concrete and construction sectors, which use natural resources for manufacturing. It also gives alarming notes on sustainability objectives. Since ancient times, the use of lime-based building materials has been practised. The lime mortar was prepared by mixing different natural additives (such as molasses, curdling, and egg yolk), which help in gaining strength as the lime mortar has less strength than the cement mortar, but it is more durable than the cement mortar. Many heritage monuments of the Indus Valley civilization made of lime mortar attest to the durability of lime-based structures. Lime-based mortar is still used in rural and suburban areas of India, along with other locally viable reinforcing materials like rice husk and tree barks. Although the prepared building material is eco-friendly enough, it still has a higher dead weight and some durability issues (e.g. higher water absorption). Here, we discuss the surface treatment of the waste polyethylene bag cuts, which have a higher potential for littering, using the traditional building material lime (usually used as the main binding material). Thus, lime-coated polyethylene bag cuts were prepared. When those lime-coated polymeric wastes were employed in wet, aerated cement-based slurry, they offered higher performance in fibre-matrix bonding than that of the untreated polyethylene bag cuts. Additionally, we used an eco-friendly aeration technique that required lesser energy than the presently available aeration techniques (i.e., an aluminium powder-based method). Colloidal gas aphrons (CGAs), which are micron-sized bubbles, are introduced as an aerating medium in this novel technique. The resultant aerated blocks possess low densities, moderate [1.95â€“3 MPa] compressive strengths, absorb less water, and are less prone to cracking due to the bridging effect of the plastic bag cuts. We were successful in using fly ash, another waste, during the preparation of the aerated slurry, which reduces the need for cement, main binder, and lowers the carbon impact. Thus, by using waste materials, these innovative reinforced aerated blocks would be able to achieve sustainability while also improving the characteristics of aerated blocks.

Reducing Water Use By Increasing Cooling Tower Cycles of Concentration

Authors

Keister, T. - Presenter

Reducing Water Use by Increasing Cooling Tower Cycles of Concentration

Abstract: Water is the cooling medium in many industrial processes and for air conditioning in commercial and institutional buildings. The majority of this water is recirculated via evaporative cooling towers. In many facilities evaporative cooling towers consume up to 35% of total input water. Part of the water used is removed from the cooling tower system as blowdown to prevent scale formation. Blowdown amount is controlled by the cycles of concentration (COC) required to prevent scale formation. The majority of cooling towers are supplied with hard makeup water which typically results in a COC between 2 and 4. Increasing COC to 8 to 10, or more, substantially reduces blowdown and thus fresh water use.

Reported will be the use of softened makeup water to increase cooling tower COC to a minimum of 8. Concentrated soft water is very corrosive and the patented water treatment chemistry invented to control corrosion in cooling tower systems operated at high COC will be discussed. Case history data will be present on several cooling tower systems converted to use of softened makeup water. Water usage, analysis results, and corrosion coupon data will be presented along an economic evaluation of the process.

Embracing the Challenges of Sustainability to Help Feed a Growing World

Authors

Kumar, S. - Presenter, ADM

ADM, a global leader in human and animal nutrition, unlocks the power of nature to enrich the quality of life. Because the company occupies a unique position in the global food and agriculture value chain extending from the farm gate to the consumerâ€™s plate, it is possible to drive transformative changes to help strengthen the integrity of global supply chains, and to help decarbonize the industry to safeguard the planetâ€™s future.

ADM is expanding its sustainability efforts to meet the needs of the growing global population, and to give customers the edge they need to navigate new consumer demands, shifting attitudes, and environmental challenges. As part of these efforts, the company continues to set ambitious sustainability goals for itself, including:

- The â€œStrive 35â€? commitments to reduce absolute greenhouse gas emissions by 25%, energy intensity by 15%, water intensity by 10%, and achieve a 90% landfill diversion rate by 2035.
- An aspiration to achieve net-zero status by 2050.
- A commitment to aligning our sustainability goals with the Science-Based Target Initiative for limiting the average rise in global temperature to 1.5 degrees Celsius.
- A commitment to ensuring that low-carbon energy represents 25% of our total energy usage by 2035.

Among ADMâ€™s most notable programs are those aimed at expanding regenerative agriculture, enhancing process efficiencies, expanding carbon capture at additional U.S. facilities, expanding the use of biofuels in transportation fleets, and replacing coal-fired boilers. This presentation provides an overview of many of these initiatives.

Mobilizing the Chemical Industry to Decarbonize Energy: A Material Science Company's Role and Examples

Authors

Lemonds, A. - Presenter, Dow

Today, technological solutions at commercial scale to lower the energy sectorâ€™s carbon emissions to the atmosphere exist. Many of these solutions are chemistry-based, and the chemical industry should be ready to implement such solutions when regulatory and socioeconomic environments open windows of opportunity. Furthermore, the International Energy Agencyâ€™s *Net Zero by 2050* projects a need to capture and sequester 7.6 Gt/a of CO2. This is a phenomenal amount of CO_2 to be addressed, which might be best characterized as a waste management challenge. Capture, by any means, adds cost, and sequestration solutions have advanced further than feasible uses for CO_2 volumes of this magnitude. Therefore, it is imperative that the chemical industry fully appreciates the magnitude and challenge of carbon removal before it to achieve such goals. This presentation will share examples of The Dow Chemical Companyâ€™s journey to bring to the market offerings for decarbonization and to decarbonize its own manufacturing footprint. The term "decarbonize,â€� for the sake of this report, will refer to measures that lower CO_2 emissions to the atmosphere. Related learning will point to the importance of collaborations across the value chain â€" with technology licensors, engineering companies, material suppliers, and operating companies. Moreover, the report will discuss the importance of supply chain resiliency, performance guarantees, and a deep understanding of end use applications to back the substantial capital required. Regarding process chemicals that may be used, specific examples will include chemical and physical solvents for pre- and post-combustion CO_2 capture and for renewable fuels processing, surfactants for enhanced oil recovery, and heat transfer fluids for concentrated solar power generation. For decarbonizing chemical manufacturing, examples will include Dowâ€™s plans for net-zero ethylene production and small modular nuclear reactors for co-generation.

Study on Ion Batteries and Soc Estimation

Authors

Chaware, M. - Presenter, Vishwakarma Institute of Technology, Pune

More, S. - Presenter, Vishwakarma Institute of Technology, Pune

Chandwani, N. - Presenter, Vishwakarma Institute of Technology, Pune

Gaikwad, A., Vishwakarma Institute of Technology, Pune

Bhokare, S., Vishwakarma Institute of Technology, Pune

Misal, D., Vishwakarma Institute of Technology, Pune

This study compares lithium-ion (LIB) and aluminium-ion (AIB) batteries. While LIBs have been dominant, AIBs are gaining attention for their resource availability and potential for high energy density. The study evaluates performance metrics, materials, and manufacturing processes for both battery types. It finds that LIBs currently have higher energy density and longer cycle life, but AIBs offer advantages in resource availability, cost-effectiveness, and environmental impact. The study discusses recent advancements and ongoing research, providing valuable insights for researchers, industry professionals, and policymakers in the pursuit of efficient and sustainable energy storage solutions.

This review provides an overview of techniques for estimating the state-of-charge (SOC) and state-of-health (SOH) of batteries. It discusses methods for accurately determining SOC, including coulomb counting and model-based approaches. Various approaches for assessing SOH, such as capacity fade analysis and impedance spectroscopy, are also covered. The review highlights the importance of precise SOC and SOH estimation in applications like electric vehicles and renewable energy systems for optimal battery management. The findings assist researchers and engineers in selecting suitable monitoring methods, contributing to the development of more reliable and efficient energy storage systems.

Optimization for Sustainable Process Design

Authors

Tabat, M. - Presenter

Kibria, M. G., University of Calgary

McCoy, S., University of Calgary

Epelle, E. I., University of Edinburgh

One important way to support the global goal of net-zero emissions is the development of processes that reduce environmental impacts, including CO_2 emissions. This is the domain of process systems engineering (PSE). However, achieving this goal requires the development of PSE tools that focus not only on the optimization of technical and economic performance but the incorporation of environmental performance in process synthesis. Such optimization and decision support tools must be sufficiently robust that can support the exploration and analysis of various process alternatives under uncertainty and produce an optimal balance between profit maximization, energy efficiency, and environmental performance. This paper focuses on developing PSE tools that allow for the synthesis and optimization of processes that meet environmental design goals along with technical and economic performance.

We formulated a multiobjective optimization in MATLAB® in which the objective functions are minimizing total cost and carbon footprint subject to certain constraints while the process alternatives are modelled as decision (binary) variables. Aspen Plus® Software is utilized to model these alternative processes with accurate mass and energy balances. The results from the simulation were regressed to formulate mathematical models for the optimization.

A genetic algorithm solver is utilized to generate a Pareto front which shows the trade-off between carbon footprint and total cost, at the same time, giving the optimum reactor configuration at each local minimum. The multiobjective optimization using the stochastic base approach is compared with the gradient-based approach.

These tools will be presented in the specific context of syngas production from municipal solid waste (MSW) via steam gasification. Four different reactor configurations namely, plasma gasifier, entrained flow reactor, fluidized bed reactor, and fixed bed reactor to produced syngas were modelled and optimal configurations at best operating conditions were obtained with the constraint that the molar ratio of hydrogen to carbon monoxide (H2:CO) is suitable for the Fischer Tropsch process.

The results provide a set of optimal process alternatives and insight for decision-makers and companies to decide depending on their priorities. Similarly, the outcome of this work will demonstrate the application of superstructure optimization in process design and its effectiveness in decision-making.

Sustainability Assessment of Purpose Grown Sustainable Aviation Fuel Feedstocks in the Contiguous United States

Authors

Limb, B. - Presenter, Colorado State University

Quinn, J., Colorado State University

Simske, S., Colorado State University

Smith, J., B&D Engineering and Consulting

The US governmentâ€™s Sustainable Aviation Fuel (SAF) Grand Challenge strives to increase annual SAF production to 35 billion gallons by 2050 while decreasing aviation emissions by 50%. This goal has increased domestic interest in purpose-grown bioenergy crops (*Miscanthus*, algae, etc.); however, limited research exists on the large-scale, country wide feasibility of SAF feedstocks in the US. Therefore, this research evaluates the county-level bioenergy production potential, economics, and environmental impact of SAF feedstocks throughout the Contiguous United States (CONUS) using GIS analysis and multi-objective optimization.

In total, six bioenergy crops (switchgrass, *Miscanthus*, poplar, algae, corn, and soybeans) and eight land types (barren, deciduous forest, evergreen forest, mixed forest, shrubland, grassland, pastureland, and cropland currently used for bioenergy) were considered for bioenergy production potential. The eight land classifications considered represent 71% of total CONUS land area. GIS mapping techniques were used to partition the CONUS into 200 m by 200 m patches to understand bioenergy production feasibility on various land types while excluding land with low yield, unfavorable terrain (e.g. steep slopes), or a protected status. County-level results include bioenergy land availability, biofuel production potential, minimum fuel selling price, and greenhouse gas emission impacts including direct land use change (dLUC).

Results show that *Miscanthus* is the most promising bioenergy crop by providing both a competitive fuel selling price ($4.49-$5.18/gal) and low emissions (-0.5â€“3.8 kg $CO_{2\text{-}eq}$/gal vs 10.7 kg $CO_{2â€‘eq}$/gal for conventional jet fuel). All feedstocks grown on forestland were found to have a larger environmental impact than conventional jet fuel due to dLUC impacts. Algae was found to have the highest fuel selling price ($9.98-$12.51/gal) and emissions (8.2-9.9 kg $CO_{2â€‘eq}$/gal) but it also has a biofuel productivity (fuel per land area) 10X that of the other feedstocks. Multi-objective optimization results illustrate the tradeoffs between optimizing CONUS SAF production for minimum fuel selling price, emissions, or land use. When minimizing for emissions, 13.4% of total CONUS land is required and *Miscanthus* is the primary feedstock. However, when land area is minimized only 0.8% of CONUS land is required because high productivity algae is deployed as the feedstock.

Production of Sustainable Hydrogen from Lignocellulosic Biomass: Towards Integrated Biorefinery

Authors

Elsaddik, M. - Presenter

Nzihou, A., Mines Albi, CNRS, Centre RAPSODEE, Univ. Toulouse

According to the *Net Zero Emissions by 2050* (NZE2050) scenario of the International Energy Agency (IEA), the global CO2 emissions need to be reduced by around 45% from 2010 levels by 2030 and reach net zero by 2050. In such a scenario, hydrogen is projected as a fascinating energy carrier among the environmentally friendly solutions by using sustainable resources for its production. However, the production of hydrogen from renewable resources towards a sustainable circular bioeconomy is rare to date. In this framework, the production of H_2-rich syngas from an organosolv cellulose pulp is presented.

In the first step, the softwood sawdust is fractionated on a laboratory scale into its main components (cellulose, lignin, and hemicellulose) using a mild-conditions formic acid organosolv process. The fractionation was subsequently upscaled from lab scale to pilot scale. Silicon (Si) was the main inherent mineral element in the resulting cellulose pulp while alkali metals were dissolved.

Next, different experimental conditions were applied to carry out steam gasification of the cellulose pulp in a lab-scale semi-continuous reactor using a in a multi-stage process approach: a slow pyrolysis followed by the steam gasification of biochar. At temperatures below 950°C, the steam gasification of cellulose pulp biochar was significantly hindered by the inhibitory effect of Si.

At 950°C, the results showed a great potential hydrogen production up to 56.3 vol.% and 40 $g_{H2}/kg_{cellulose\ pulp}$. The process was modeled to examine the overall performance and evaluate the heat integration. When integrating organosolv treatment to steam gasification, hydrogen production efficiency dropped from 41.5 to 26.2% while the energy requirement raised from 77 to 111 kWh/kg (H_2).

Further assessment in terms of solvent and by-products recovery is necessary to improve process efficiency. Thus, hydrogen from integrated biorefineries could potentially complement electrolysis (51%, 50-65 kWh/$_{kgH2}$).

The ability to generate hydrogen (H2) through cellulose pulp and create value-added products using lignin and hemicellulose opens up fresh possibilities for the establishment of environmentally friendly biorefineries that operate on a closed carbon cycle. These biorefineries can be developed using biomass sources unrelated to food production, contributing to sustainable and renewable energy solutions.

From setting Net-Zero targets to implementation – a global perspective

Authors

Prox, M. - Presenter, iPoint Systems gmbh

Incentives and pressure from investors and regulators have continued to increase since 2020 to more and more net zero commitments of industry-leading companies across all sectors. Establishing baselines for corporate GHG Accounting has been the focus in recent years and now the attention is on implementation. Since 60-90% of the potential reduction for Green House Gases can only realized together with the supplier-network, the product-perspective is crucial to substantiate commitments and the road towards net zero. This leads to two necessities: first scaling product carbon footprints (PCF) to all products within multinational companies, and second to increase the use of supplier specific PCFs over time to ensure that GHG reduction monitoring becomes based on supply chain specific data where it is material. This means replacing industry-average data, which is always used in the first iterations to assess and calculate PCFs, with supplier-specific data.

The keynote will provide an overview of industry initiatives on harmonizing calculation methods and PCF sharing and the implications for organizations that belong to or deliver into various sectors. Initiatives taken into account include Together for Sustainability, Catena-X, Pathfinder by WBCSD. The presentation will furthermore give insights into the journey towards automated PCFs and Life Cycle Assessment (LCAs) from organization of the sectors Chemical, Automotive and Electronic. The audience will understand how IT solutions as well as internal and external collaboration are used as an enabler to address the challenges of complexity in products, supply networks and/or highly inter-connected global production systems, when scaling PCFs.

Phosphate Modified Baker's Yeast: An Effective Approach for Simultaneous Lead Removal and Water Hardness Reduction in Wastewater

Authors

Zamanova, A. - Presenter

Rahimli, G. - Presenter, Baku Higher Oil School

Guliyeva, R., Baku Higher Oil School

Throughout this scrutiny, the efficacy of employing phosphate-modified baker's yeast (PMBY) as a bio-sorbent for the bioremediation method for lead (Pb) removal and the reduction of water hardness in wastewater was investigated. Overall, the adsorption method of lead from wastewater was based on the ion exchange method as well as chelation as phosphorous and oxygen atoms possess a strong affinity for metal ions such as lead. Because of lead's detrimental impact on human health and the environment, lead pollution is a consequential environmental issue. In the yeast modification process, potassium dihydrogen phosphate and potassium hydrogen are used as phosphate sources, while other materials include urea, baker's yeast, and lead nitrate as lead sources for wastewater samples. Devices for the experiment were a shaker (SK-O330-Pro), centrifuge (EPPENDORF Centrifuge 5430), vacuum dryer (BINDER VDL), vacuum filtration (NALGENE Rapid-Flow), and grinder; furthermore, the analysis of three treated wastewater samples, with 0.05 grams, 0.1 grams, and 0.2 grams of the modified yeast adsorbent, was conducted with a microwave plasma atomic emission spectrometer (Agilent Technologies 4200 MP-AES) for the determination of elements. The initial concentration of wastewater was 50 ppm. The results showed that the lead concentration decreased to 2.06 ppm, 0.31 ppm, and 0.26 ppm, respectively. Although it was not our purpose to focus on reducing the hardness of water, we discovered that, with phosphate-modified yeast, we decreased the amount of calcium and magnesium inside the water. Thus, the amounts of calcium and magnesium in the water reduced, which reduced water hardness, with the numbers decreasing from 1.932 ppm to 1.069 ppm and from 1.3586 ppm to 0.7285 ppm, respectively. The amount of lead in the water significantly decreased as the adsorbent concentration increased from 0.05 to 0.2 with the removal efficiency increasing from 96% to 99.5%. Through the experimentation of a sustainable bioremediation strategy for dealing with lead contamination and water hardness issues in wastewater, this research contributes to the field of environmental remediation. The outcomes demonstrate how successfully modified yeast can be used as a practical water treatment method.

Valorizing CO_2 using renewable energy to enable the future of food

Authors

Dunn, C. - Presenter, Circe Bioscience

Circe is creating the future of fermentation to make use of waste CO_2 to produce high value goods and enable a circular carbon economy. While much of the focus is on how to sequester as much carbon as posible, we also need to find ways to reduce our CO_2 output while still supplying the products demanded by consumers. Through the use of novel synthetic biology and gas fermentation Circe is able to transform a variety of CO_2 streams into useful materials, including fat for food, personal care and more.

Sustainability Concepts in the Food-Energy-Water Nexus

Authors

Das, T. K. - Presenter, Saint Martin's University

The idea of sustainability emerged from the World Commission on Environment and Development. Environmental sustainability and/or sustainable development â€œoccurs when we maintain or improve the material and social conditions for human health and the environment over time without exceeding the ecological capabilities that support themâ€�. As such, sustainability has three underlying bases: (1) economic viability - the cost and business aspects of a project; (2) social concern - human health and social welfare; and (3) natural or ecological issues - depletion of natural capital and environment. Sustainable development has become a major driving initiative in engineering businesses throughout the world today. Our focus is on all aspects of sustainable development including environmental, industrial, business, economical, ethical, and social sustainability as they impact the availability of food, energy, and water. This is important because it has been generally acknowledged that the increasing use of non-renewable resources to support an increasing population has created an unsustainable or nearly unsustainable situation where the ability to provide adequate food, energy, and water is in question. Should this rate of use be maintained, two dire consequences will follow: first, the future generations will be unable to maintain a good standard of living, and second, developing countries will have less of an opportunity to bring their living standards to a level comparable to that of the affluent countries. This is important because the availability of food, energy, and water is an indispensable necessity for maintaining a good standard of living. The practice of chemical engineering, perhaps more than any other technical discipline, involves the use of both non-renewable and renewable materials, and energy and exergy (available energy) resources for the production of chemicals, value-added goods, services of commerce, and waste as pollution. This paper primarily deals with the ideas of environmental sustainability, methodologies to redesign sustainable chemical and allied processes and products development using the LCA approach. A case study for the production of ethanol, an energy source, from molasses, a human food, in a combined agricultural and chemical process that requires substantial amounts of water.

Effects of Organic Pollutants on Struvite Crystallization for Phosphorus Recovery from Wastewater and the Molecular Mechanism of Inhibition on Crystal Growth

Authors

Yao, M. - Presenter

Chen, M., Tianjin University

Gong, J., Tianjin University

The abilities of recovering phosphorus (P) resources and reducing water eutrophication make struvite crystallization technology attract increasing interest in wastewater treatment, but struvite crystallization process may be affected by various impurities in wastewater. In this work, the effects of nine representative ionic surfactants including three types (anionic, cationic and zwitterionic) on crystallization kinetics and product quality of struvite were investigated, and the influencing mechanism was further probed. The results demonstrated that anionic surfactants significantly inhibit crystal growth so as to reduce crystal size especially in *a*-axis direction, change crystal morphology and decrease P recovery efficiency, and also lead to a slight decline in product purity. In contrast, cationic and zwitterionic surfactants have no obvious influence on the formation of struvite. A series of experimental characterizations and molecular simulations collectively revealed that the inhibition of crystal growth by anionic surfactants are attributed to the adsorption of anionic surfactant molecules on struvite crystal surface and subsequent blockage of active growth sites. The binding ability of surfactant molecules with the Mg^{2+} exposed on struvite crystal surface was highlighted to be the most essential factor determining the adsorption behavior and adsorption capacity. Anionic surfactants with stronger binding ability with Mg^{2+} have more intense inhibitory effect, but a large molecular volume of anionic surfactants will weaken the adsorption capacity on crystal surface so as to reduce the inhibitory effect. Contrastively, cationic and zwitterion surfactants without binding ability with Mg^{2+} have no inhibitory effect. These findings enable us to have a clearer understanding of the impact of organic pollutants on struvite crystallization and make a preliminary judgment on the organic pollutants that may have the ability to inhibit the crystal growth of struvite, facilitating the implementation of struvite crystallization technology for phosphorus recovery from wastewater.

Development and Upscaling of Novel Hollow Fibre Membranes for Post-Combustion Flue Gas CO2 Separation and Capture in Gas-Fired Power Plants and Other Industries.

Authors

Low, M. - Presenter

Carbon capture is a crucial technology for reducing greenhouse gas emissions from industrial processes. One promising method for capturing carbon dioxide (CO_2) is through the use of selective membrane technology, which separates CO_2 from flue gas based on its size and charge.

Hollow fibre membranes are a type of selective membrane that have shown promise for post-combustion flue gas carbon capture. These membranes are made up of thousands of tiny hollow fibres, which are more permeable to certain gases such as CO_2, but less to others like nitrogen (N_2). When flue gas is passed through the membrane, the CO_2 selectively diffuses through the fibres and can be captured for storage or use.

While hollow fibre membrane technology has shown great potential for carbon capture, scaling up this technology from laboratory-scale experiments to large-scale industrial applications presents several challenges. These challenges include the need for highly selective and durable membranes, efficient system design, and cost-effective operation.

In this on-going study, we aim to develop (1) high performance polymeric membranes to capture CO_2 from power plants at near ambient pressures (i.e., 1 - 2 bar) and (2) scale up and commercialize it to capture CO_2 from Singapore power plants, petrochemical and industrial plants. The newly developed hollow fiber membranes and pilot systems would ideally have a CO_2 permeance of 2000-3000 GPU and a CO_2/N_2 selectivity of 30-35.

The project not only integrates fundamental membrane studies with multi-disciplinary and multi-faceted experimental investigations but also includes scaling up, business development and commercialization. The success of this program will have an immediate and significant impact on the minimization of CO_2 emission in Singapore and bring the world to be more eco-friendly.

Biomass-Derived Blue Pigment (Indigoidine) Could be Less Costly and Have Lower Carbon Footprint Than Synthetic Indigo

Authors

Baral, N. - Presenter, Lawrence Berkeley National Laboratory

Banerjee, D., Lawrence Berkeley National Lab

Eng, T., Joint BioEnergy institute

Mukhopadhyay, A., Lawrence Berkeley National Laboratory

Scown, C. D., Lawrence Berkeley National Laboratory

The current industrial production of indigo dye involves the use of toxic and hazardous compounds, resulting in the release of hazardous wastes and retention of toxic impurities in the indigo-dyed denim. To address these health and environmental concerns, and to meet the increasing demand for blue dye at a competitive price, we present a novel biologically-derived alternative to synthetic indigo dye called indigoidine. This alternative is produced using bioenergy sorghum feedstock and an ionic liquid-based one-pot high-gravity bioconversion process. We provide a comprehensive techno-economic analysis and lifecycle assessment of the biomass-derived indigoidine, which identifies areas for process improvement, cost reduction, and establishes performance targets for future research and development.

Currently, the yield of indigoidine in the one-pot batch process results in a high production cost and GHG emissions of \$29.5/kg and 25.6 $kgCO_{2e}$/kg-indigoidine, respectively. We find that key cost and carbon footprint drivers for the biomass-derived indigoidine include the indigoidine yield from sugar and lignin monomers, the solid loading rate for bioconversion, the quality of biomass feedstocks, the retention time and aeration rate for the bioreactor, sugar yield, and ionic liquid recovery. By optimizing these parameters to near theoretical limits, we were able to achieve a significant reduction in both selling price and GHG emissions, resulting in a selling price and GHG emissions of \$0.86/kg and 0.93 $kgCO_{2e}$/kg, respectively, representing a six- and five-fold reduction relative to synthetic indigo dye.

With further research and development efforts aimed at achieving the identified targets, we estimate a 95% likelihood of achieving a selling price and GHG emissions of indigoidine of less than \$1.6/kg and 2.2 $kgCO_{2e}$/kg, respectively. These values are lower than the current market price and GHG emissions of synthetic indigo dye, which are \$5.51/kg and 4.72 $kgCO_{2e}$/kg-indigoidine, respectively. Our findings suggest that the biomass-derived indigoidine could replace the fossil counterpart, saving several million dollars.

Uranium Extraction; Uranyl Ion; Seawater; Adsorption Polymer; Adsorptionkinetics; Separation; Reusability; Wind Turbine; Buoyant Platform; Chelating Ligand.

Authors

Gyasi, S. - Presenter

Technically, it is possible to separate uranium from seawater. This design paper seeks to propose a process intensified way to separate uranium from seawater using amidoxime containing polymer due to the high affinity of the uranylâ€"carbonate complex towards the polyamidoxime resin. The process justification was based on the economic criteria: production cost, physicochemical criteria: concentration and reactivity, technical criteria: extracting methods from a complex aqueous system. The separation process firstly is the adsorption of uranyl ions using ionic exchange resin in the form of a chelating polymer adsorbent. A buoyant platform with adsorbent polymer made of polyethylene fibres that comes into contact directly with sea water powered by a wind turbine which is mounted on top of the platform, continuously takes the adsorbent from the seawater after campaign time through an elution and regeneration process and then returns it to the seawater for reuse. After the adsorbent has been loaded with uranyl ions, additional downstream processes is explored to recover the uranium including cleansing of the adsorption polymer to remove organic materials, desorption: elution of the adsorbed uranium ions from the adsorption polymer with a suitable solution, purification of the eluent: removal of other desorbed compounds, concentrating the solution, solvent extraction of uranium from the solution with a mixture of an organic solvent and a specific complexing agent and finally conversion into yellow cake or uranium oxide ð'ˆ308. Next, factors that goes into the Uranium separation cost and financial justifications have been included. In future, we recommend researchers to make use of organically sourced, biodegradable substrates such as chitin which could be a good source of green chemistry alternative to conventional plastic polymers that represents a potential source of an uneasy degradable solid pollutant.

Electro-Bio Conversion for Renewable Materials, Fuels and Chemicals

Authors

Yuan, J. - Presenter

Zhang, P., University of Cincinnati

Chen, K., Texas A&M University

Xu, B., Texas A&M University

Dai, S., Texas A&M University

Recent breakthroughs highlighted the possibility to substantially improve CO_2 conversion rate and efficiency via integrating ectrocatalytic CO_2RR (CO_2 reduction reaction) to C2+ intermediates and bioconversion to diverse products. Traditional electrocatalytic CO_2RR can achieve high efficiency, yet is limited in product profile, producing primarily C1 or C2 products. Integrating catalytic CO_2 reduction with bioconversion could substantially advance the product profile to produce chemicals, fuels, materials, and food to replace the current petrochemical and agriculture products. The new electro-bio systems could substantially improve the solar-to-product efficiency, advance carbon capture and utilization, and mitigate climate change. However, the state-of-the-art electro-bio systems are limited by inefficient electron and mass transfers, unfavorable metabolic kinetics, and inadequate molecular building blocks. We overcome these barriers with the systematic design of electrocatalysis, the chem-bio interface, and microorganisms to enable efficient electro-microbial conversion with C2 (EMC2) intermediates. The soluble C2 intermediates can facilitate rapid mass transfer, readily enter primary metabolism, have less toxicity, carry more energy and electrons, and serve as better molecular building blocks for many microorganisms. Despite the advantages, the integration of chemical and biological conversions must overcome the intermediate incompatibility, the harsh chemical catalysis conditions, and the inefficient mass, energy, and electron transfers. On one hand, we have designed the electrolytes, electrode, electrolyzer, and catalyst configurations to allow efficient CO_2RR into acetate and ethanol at conditions compatible to biological conversion. On the other hand, we hereby carried out multi-module cellular design to enhance bioproductivity, facilitate efficient substrate usage, and improve reductant production. Furthermore, the systematic design achieved integrated, continuous, and rapid microbial biomass and PHA production from CO2 with record-level productivity. The multi-layer EMC2 design has achieved 6 and 8 times increase of microbial biomass productivity compared to C1 intermediate and hydrogen-driven routes, respectively. Moreover, the systematic design achieved integrated, continuous, and rapid microbial biomass and PHA production from CO2 with record-level productivity. Overall, the EMC2 system could achieve 4.5% solar-energy-driven CO_2 conversion to biomass, which is multiple times higher than the natural photosynthesis. Further electrocatalytic and microbial design has enabled novel routes for the efficient conversion of CO_2 into various food, chemicals, and fuels.

Challenges in Reallizing Large Scale Carbon Capture and Storage (CCS) Solution

Authors

Hashim, W. - Presenter

Carbon Capture and Storage (CCS) is a key enabler for carbon emission reduction towards a sustainable future while fulfilling energy demand. However, CCS, especially when implemented would be a great cost to plant or development project. The success of CCS projects would lie upon the availability of suitable storage sites, prioritisation of emission sources to be captured, capture technology operating condition and the power availability for the CCS facilities in order to achieve net negative carbon emission. Collaboration of surface and subsurface engineers is required in identifying and screening of CO2 storage location depending on storage capacity, integrity, and containment of the injected CO2 over time. The storage capacity would also be determined by the sources to be captured by the facilities. In some cases especially with high CO2 natural gas sources from the reservoir, the sources would be less challenging to be captured compared to the post combustion with lower CO2 concentration. The capture process of these two sources differs significantly and this would effect not just the economics but also the power consumption of the facilities. Comprehensive technology screening is also required in order to optimised the energy while capturing the CO2 emission for storage. For some capture technology, the hydrocarbon and contaminants captured would require to be recovered and removed from the injection stream. Capturing CO2 from production or processing facilities is energy intensive and would contribute to the emission of the CCS facilities. Therefore, selection of the power sources would determine the net emission reduced for a facilities. All these factors would need to be considered while evaluating the CCS design and project decision as it will ultimately affects the cost. Stakeholder management plan inclusive of technology providers and CCS experts, commercial and Health, Safety and Environment (HSE) was also need to be considered in reallizing a large scale Carbon Capture and Storage (CCS) solution. Evaluation of all these factors would enabled critical decisions and proposals to be properly assessed to realize this solution.

Water in the Emerging Energy Environment and Its Sustainable Use

Authors

Dey, H. - Presenter

Ekmekci, B., Fluor

Water has always been intricately associated with energy. Development of any specific form of energy has required the use of water during its project cycle â€" as a feedstock (e.g. Electrolysis), for cooling, for supporting utility systems, for operation of the facilities â€" safety (firefighting, first aid with safety showers, etc). Earlier forms of energy generation were dominated by fossil, that are being phased out in favour of new sustainable forms of energy. For each of these applications water is required, the demand varying in the quantity and quality. The availability and management of the water resource plays an important part in the project type and site selection as well as operation of the constructed facility during its entire life cycle.

AWWA, in their Water 2050 Sustainability document (AWWA, September 2022) reiterated the need for reframing the value of water to reflect the need to prepare for a sustainable future. The aim will also be to achieve economies of scale of water systems through consolidation and operational efficiencies.

An important aspect of todayâ€™s industry, including design and operation of a water system is the need for sustainable solutions that satisfy the triple bottom line principles of sustainability â€" people, planet & profit. A sustainable future for water in the industry will require design of water systems through energy and water efficiency, implementation of circular principles and three Râ€™s of water â€" *Reduce, Recycle and Reuse*.

In this paper, the newer challenges faced by the water industry to satisfy the needs of the emerging energy technologies, e.g. electrolysis, Battery production are highlighted. Specific requirements for water quality and treatment options are discussed in relation to the emerging energy technologies.

AUTHOR INDEX

AIChE
120 Wall Street, FL 23
New York, NY 10005-4020

ISBN 978-1-7138-8679-2
9 781713 886792

CURRAN ASSOCIATES INC.
proceedings.com